Generalized Radon Transforms and Imaging by Scattered Particles

Broken Rays, Cones, and Stars in Tomography

Contemporary Mathematics and Its Applications: Monographs, Expositions and Lecture Notes

Print ISSN: 2591-7668
Online ISSN: 2591-7676

This series aims to inspire new curriculum and integrate current research into texts. Its aims and main scope are to publish:

- Cutting-edge Research Monographs
- Mathematical Plums
- Innovative Textbooks for capstone (special topics) undergraduate and graduate level courses
- Surveys on recent emergence of new topics in pure and applied mathematics
- Advanced undergraduate and graduate level textbooks that may initiate new directions and new courses within mathematics and applied mathematics curriculum
- Books emerging from important conferences and special occasions
- Lecture Notes on advanced topics

Monographs and textbooks on topics of interdisciplinary or cross-disciplinary interest are particularly suitable for the series.

Published

More information on this series can also be found at
https://www.worldscientific.com/series/cmameln

Contemporary Mathematics and Its Applications
Monographs, Expositions and Lecture Notes

Vol. **6**

Generalized Radon Transforms and Imaging by Scattered Particles

Broken Rays, Cones, and Stars in Tomography

Gaik Ambartsoumian

The University of Texas at Arlington, USA

W₆ World Scientific

NEW JERSEY · LONDON · SINGAPORE · BEIJING · SHANGHAI · HONG KONG · TAIPEI · CHENNAI · TOKYO

Published by

World Scientific Publishing Co. Pte. Ltd.
5 Toh Tuck Link, Singapore 596224
USA office: 27 Warren Street, Suite 401-402, Hackensack, NJ 07601
UK office: 57 Shelton Street, Covent Garden, London WC2H 9HE

Library of Congress Cataloging-in-Publication Data
Names: Ambartsoumian, Gaik, author.
Title: Generalized radon transforms and imaging by scattered particles : broken rays, cones, and
    stars in tomography / Gaik Ambartsoumian, The University of Texas at Arlington, USA.
Description: New Jersey : World Scientific, [2023] | Series: Contemporary mathematics and
    its applications : monographs, expositions and lecture notes, 2591-7668 ; vol. 6 |
    Includes bibliographical references and index.
Identifiers: LCCN 2022037836 | ISBN 9789811242434 (hardcover) |
    ISBN 9789811242441 (ebook) | ISBN 9789811242458 (ebook other)
Subjects: LCSH: Radon transforms. | Geometric tomography. |
    Scattering (Mathematics) | Tomography.
Classification: LCC QA672 .A43 2023 | DDC 515/.723--dc23/eng20221209
LC record available at https://lccn.loc.gov/2022037836

British Library Cataloguing-in-Publication Data
A catalogue record for this book is available from the British Library.

For any available supplementary material, please visit
https://www.worldscientific.com/worldscibooks/10.1142/12424#t=suppl

Desk Editors: Soundararajan Raghuraman/Lai Fun Kwong

Typeset by Stallion Press
Email: enquiries@stallionpress.com

To Ani and Sophie

Preface

Conventional tomographic modalities used in medicine (e.g. X-ray CT, positron emission tomography, and single photon emission CT) exploit primary (unscattered) radiation to acquire information about the internal structure of the body. The scattered radiation in these techniques is treated as noise that has to be filtered out or compensated for, despite the fact that scattering often accounts for the lion's share of all photon interactions with the matter. Various methods of utilizing scattered particles as sources of meaningful data for image reconstruction have been proposed and explored since the 1950s. The imaging systems designed in the early works used simplistic mathematical models, and their applications were hampered by the lack of sophisticated instrumentation and robust computational resources. The development of powerful computers and high-sensitivity detectors, including various types of Compton cameras, has renewed the enthusiasm of the scientific community for that subject. In particular, during the past two decades, there has been an explosion of interest toward mathematical models of imaging modalities using *single-scattered* radiation. Notably, new and interesting directions of research have emerged in integral geometry, prompting a need to study generalized Radon transforms integrating along nonsmooth trajectories and surfaces.

While a substantial amount of work has been done in these fields, the results have been dispersed across journal articles in various disciplines, some of which are hard to read for nonexperts. The primary goal of this book is to cover, in a detailed fashion, the relevant imaging modalities, their mathematical models, and the related generalized Radon transforms. The discussion of the latter comprises a thorough exploration of their known mathematical properties, including injectivity, inversion formulas, range description, microlocal analysis, and results of numerical simulations.

We would like to emphasize that the book is aimed at occupying a specific niche, covering integral-geometric problems related to imaging from single-scattered particles. As a result, in the treatment of generalized Radon transforms (GRT), we concentrate our attention on the operators that map a function to its integrals along curves (or over surfaces) with a singularity, which we call a "vertex." Moreover, even among such transforms, we place an emphasis on those whose "vertex" is inside the image domain, i.e. inside the support of the image function (Chapters 3–6). Operators of this type exhibit remarkably different qualitative behavior from more conventional GRTs integrating along smooth trajectories. The GRTs with a vertex outside of the image domain are typically treated by modifying the associated task to one that involves transforms integrating over smooth curves/surfaces. This is often accomplished via a continuation of the function by some sort of symmetry or replacement of the rays of integration by lines utilizing the shape of the support of the function (see discussions in Chapter 2).

We did not attempt to provide a systematic treatment of GRTs integrating over smooth curves and surfaces. To read and understand most of the material in this book, it is sufficient to know some basic results about inversion and properties of the classical Radon and X-ray transforms (e.g. see Feeman, 2015). For a more comprehensive treatment of such operators, we refer the reader to the following literature and the references therein: Ehrenpreis (2003); Gelfand *et al.* (1966, 2003); Helgason (1999, 2011); Kuchment (2013); Natterer (2001b); Natterer and Wübbeling (2001); Palamodov (2004, 2016); Rubin (2015); Volchkov (2003).

It is also important to note that we do not use or cover in any fashion imaging techniques using multiply scattered particles or scattering of acoustic and electromagnetic waves. These are extensive and important subjects in their own right, primarily based on partial differential equations and integral equations techniques, with tremendous amount of interesting results. For more information on those topics, we refer the interested reader to the following literature and the references therein: Arridge and Schotland (2009); Bal (2009); Borcea *et al.* (2002); Cakoni and Colton (2005); Cheney and Borden (2009); Colton and Kress (2013a,b); Fouque *et al.* (2007).

I express my sincere gratitude to Professor Zuhair Nashed for the offer to write this book and his constant support throughout the process of preparing the manuscript. Without his encouragement, this book would never have been published.

I am also grateful to my collaborators Rim Gouia-Zarrad, Mohammad Latifi Jebelli, Rohit Mishra, Sunghwan Moon, and Souvik Roy, the joint work with whom has served as a basis for a large portion of the material covered in the book.

I am indebted to Mark Agranovsky, Jan Boman, Raluca Felea, David Finch, Alexander Katsevich, Venky Krishnan, Peter Kuchment, Leonid Kunyansky, Matthew Lewis, Clifford Nolan, John Schotland, Todd Quinto, and others, from whom I have learned a lot about integral geometry and tomography and conversations with whom over the years have shaped my vision of the subject.

Thanks are also due to Mohammad Latifi Jebelli and Souvik Roy for providing the images of numerical reconstructions used in the book and to Rim Gouia-Zarrad and Rohit Mishra for reading various parts of the manuscript and suggesting improvements.

A portion of this book was written during a semester-long faculty development leave supported by the University of Texas at Arlington.

And last but not least, I would like to thank Rochelle Kronzek and Lai Fun Kwong of World Scientific for their patience and unconditional support during the writing of the manuscript.

About the Author

Gaik Ambartsoumian received a diploma degree in applied mathematics from the Obninsk Institute of Nuclear Power Engineering (Obninsk, Russia) in 2001 and a PhD in mathematics from Texas A&M University (College Station, TX, USA) in 2006. Since 2006, he has been on the faculty of the Department of Mathematics at the University of Texas at Arlington (Arlington, TX, USA). He has also held adjunct or visiting positions at the American University of Armenia (Yerevan, Armenia), Centre for Applicable Mathematics at Tata Institute of Fundamental Research (Bengaluru, India), UT Southwestern Medical School (Dallas, TX, USA), Mathematical Sciences Research Institute (Berkeley, CA, USA), and Applied Science Laboratory at General Electric Medical Systems (Milwaukee, WI, USA). His scientific interests include computerized tomography, integral geometry, inverse problems, and mathematical problems of imaging. His research projects have been funded by various agencies and organizations, including the National Science Foundation, the National Institutes of Health, the U.S. Department of Defense, and The Simons Foundation.

Contents

PART I
Particle Transport and Scattering Tomography

Chapter 1

Radiative Transport Equation

The radiative transport (or transfer) equation (RTE) is an integro-differential equation, which describes the transfer of particles through a medium. It serves as the backbone of the mathematical models of the imaging techniques discussed in this book.

We start this chapter with the formulation of the RTE and discuss its individual terms together with their physical meaning. Special attention is paid to the scattering term and its kernel in different setups. The second part of the chapter is dedicated to the solutions to various forms of the RTE, which lead to the consideration of an assortment of generalized Radon transforms. The list includes the divergent beam transform (also called fan-beam or cone-beam transform), attenuated ray transform, X-ray transform, and V-line transform (sometimes also called broken ray transform). The last one, together with several other generalized Radon transforms introduced in the next chapter, is the main subject of study in the second half of this book.

1.1 Formulation of Radiative Transport Equation

The propagation of radiation in a medium is accompanied by random processes of absorption and scattering of transported particles. Therefore, the mathematical models describing transport theory involve probability densities (or distribution functions) that depend on the position, direction, and energy of such particles. The detector measurements, accordingly, correspond to the mean values of the associated random variables (Barrett and Myers, 2003; Duderstadt and Martin, 1979). The imaging applications that are of interest in this book include X-ray, γ-ray, and optical radiation, where the transported particles are photons.

3

Let $w(\mathbf{r}, \hat{\mathbf{n}}, E, t)$ denote the *photon distribution function*, where $\mathbf{r} \in \mathbb{R}^3$ denotes a position vector, $\hat{\mathbf{n}} \in \mathbb{S}^2$ is a unit vector indicating a direction, E denotes the photon energy, and t is the time variable. Then,

$$N_a = w(\mathbf{r}, \hat{\mathbf{n}}, E, t) \, dV \, d\Omega \, dE \qquad (1.1)$$

is the average number of photons at a moment of time t contained in a volume dV centered at point $\mathbf{r}$ and traveling in a solid angle $d\Omega$ around direction $\hat{\mathbf{n}}$ with energy between E and $E + dE$.

The *RTE*, also often called *Boltzmann equation*, is an integro-differential equation describing particle transport in the form of a balance relation (Barrett *et al.*, 1999). In essence, it states that the intensity of radiation at a given location changes with time due to absorption, scattering, emission, and propagation of particles (see Figure 1.1). Namely,

$$\frac{\partial w}{\partial t} = \left[\frac{\partial w}{\partial t}\right]_{\mathrm{ab}} + \left[\frac{\partial w}{\partial t}\right]_{\mathrm{sc}} + \left[\frac{\partial w}{\partial t}\right]_{\mathrm{em}} + \left[\frac{\partial w}{\partial t}\right]_{\mathrm{prop}}. \qquad (1.2)$$

During absorption, an incident photon disappears by transferring all of its energy to the interacting atom of the medium. The term scattering stands for a variety of phenomena where an incident photon interacts with the medium and changes its traveling direction. Emission means generation of photons by radiation sources. Finally, propagation represents the transfer of moving photons without any interaction with the medium.

Let us now briefly discuss each of the four terms in (1.2). For a detailed derivation of the corresponding relations, see Barrett and Myers (2003).

The absorption component of the time derivative satisfies the following relation:

$$\left[\frac{\partial w}{\partial t}\right]_{\mathrm{ab}} = -c\,\mu_a(\mathbf{r}, E, t)\, w(\mathbf{r}, \hat{\mathbf{n}}, E, t), \qquad (1.3)$$

where c is the speed of light in the medium and $\mu_a(\mathbf{r}, E, t)$ is the *absorption coefficient* of the medium.

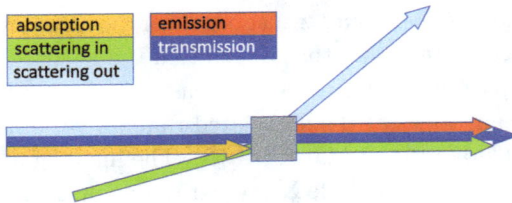

Fig. 1.1   A sketch of various physical interactions described by RTE.

In general, the speed of light can also depend on the position in the medium, energy of the radiation, and time. Moreover, in an anisotropic medium, it can also depend on the direction of propagation. However, here and in the rest of the book, we assume that the medium is homogeneous, i.e. the speed of light c is constant.

The emission of particles is described by the *source density function* $q(\mathbf{r}, \hat{\mathbf{n}}, E, t)$ as follows:

$$\left[\frac{\partial w}{\partial t}\right]_{\text{em}} = q(\mathbf{r}, \hat{\mathbf{n}}, E, t). \tag{1.4}$$

If the emission happens at near constant energy levels $E = E_0$, then the right-hand side of (1.4) can be substituted by $q(\mathbf{r}, \hat{\mathbf{n}}, t)\, \delta(E - E_0)$, where $\delta(\cdot)$ denotes the Dirac delta function. In many applications, radioactive sources emit isotropically, in which case the source density function is independent of $\hat{\mathbf{n}}$.

The propagation component of the time derivative of photon distribution function can be expressed as follows:

$$\left[\frac{\partial w}{\partial t}\right]_{\text{prop}} = -c\,\hat{\mathbf{n}} \cdot \nabla w(\mathbf{r}, \hat{\mathbf{n}}, E, t). \tag{1.5}$$

The scattering component of the time derivative is more complicated. Since it plays a prominent role in the derivation of the integral-geometric models studied in this book, it is worth analyzing $[\partial w/\partial t]_{\text{sc}}$ in greater detail.

Scattering has two effects on a group of photons in a small volume dV traveling in a small solid angle $d\Omega$ with an energy between E and $E + dE$. The average number of photons in that group is expressed by formula (1.1). On the one hand, scattering can decrease that number since some photons from the group may move out of it due to change in direction, energy, or both. On the other hand, certain photons that were not in that group may move into it by scattering into the solid angle $d\Omega$ and the energy range $(E, E + dE)$. We consider each of these components of scattering separately:

$$\left[\frac{\partial w}{\partial t}\right]_{\text{sc}} = \left[\frac{\partial w}{\partial t}\right]_{\text{in}} + \left[\frac{\partial w}{\partial t}\right]_{\text{out}}. \tag{1.6}$$

Scattering out of the group is analogous to absorption and is expressed in a similar fashion:

$$\left[\frac{\partial w}{\partial t}\right]_{\text{out}} = -c\,\mu_s(\mathbf{r}, E, t)\, w(\mathbf{r}, \hat{\mathbf{n}}, E, t), \tag{1.7}$$

where $\mu_s(\mathbf{r}, E, t)$ is the *scattering coefficient* of the medium.

Scattering into the group is more elaborate since one has to take into account (i.e. integrate over the set of) all traveling directions and all possible energies. Namely,

$$\left[\frac{\partial w}{\partial t}\right]_{\text{in}} = \mathcal{K}w \doteq \int_{4\pi} \int_0^\infty K(\hat{\mathbf{n}}, E; \hat{\mathbf{n}}', E'|\mathbf{r}, t)\, w(\mathbf{r}, \hat{\mathbf{n}}', E', t)\, dE'\, d\Omega', \quad (1.8)$$

where the solid angle of 4π steradians corresponds to the entire unit sphere $\mathbb{S}^2$ in $\mathbb{R}^3$. The kernel $K(\hat{\mathbf{n}}, E; \hat{\mathbf{n}}', E'|\mathbf{r}, t)$ describes the conditional probability of a particle changing its original direction and energy $(\hat{\mathbf{n}}', E')$ to $(\hat{\mathbf{n}}, E)$ after scattering at a given location and time $(\hat{\mathbf{r}}, t)$. In isotropic media, the dependence of kernel K on directions $\hat{\mathbf{n}}$ and $\hat{\mathbf{n}}'$ can be viewed simply as a dependence on the scattering angle θ_s or, equivalently, on the inner product $\hat{\mathbf{n}} \cdot \hat{\mathbf{n}}' = \cos\theta_s$.

Combining equations (1.2)–(1.8), we get the complete form of the RTE:

$$\frac{\partial w}{\partial t} = q - c\,\mu_t w + \mathcal{K}w - c\,\hat{\mathbf{n}} \cdot \nabla w, \quad (1.9)$$

where $\mu_t = \mu_a + \mu_s$ is the *total attenuation (extinction) coefficient*.

In many applications, equation (1.9) can be simplified using some additional information about the setup. Several important examples are listed in the following sections.

1.1.1   *Time-independent setup*

In medical imaging using X-rays, γ-rays, or light, the travel time of photons through the body is on the scale of nanoseconds. If the medium and radiation sources are time independent or change slowly compared to the transit time of photons, it is reasonable to assume that $\partial w/\partial t = 0$ and the quantities involved in the RTE are independent of time. The resulting *steady-state solution of the RTE*, $w(\mathbf{r}, \hat{\mathbf{n}}, E)$, satisfies the corresponding simplified version of (1.9):

$$c\,\mu_t w + c\,\hat{\mathbf{n}} \cdot \nabla w = q + \mathcal{K}w. \quad (1.10)$$

1.1.2   *Energy-independent setup*

In cases where the predominant type of scattering is elastic (e.g. in optical imaging), particles do not lose or gain energy after the scattering event. Therefore, the corresponding kernel $K = K(\hat{\mathbf{n}}, \hat{\mathbf{n}}'|\mathbf{r}, t)$ is free of energy variables, and $\mathcal{K}w$ includes only integration with respect to the traveling directions. If, in addition, the radiation sources are monoenergetic (i.e. emit at some constant energy level E_0), then one can ignore the dependence of the quantities involved in the RTE on particle energy E altogether.

For example, in optical imaging of biological tissues with monochromatic time-independent sources, an accurate description of kernel K is given through the Henyey–Greenstein phase function (Henyey and Greenstein, 1940; Jha *et al.*, 2012) as follows:

$$K(\hat{\mathbf{n}}, \hat{\mathbf{n}}'|\mathbf{r}) = \frac{c\,\mu_s(\mathbf{r})}{4\pi} \frac{1 - g^2}{\left[1 + g^2 - 2g\,\hat{\mathbf{n}} \cdot \hat{\mathbf{n}}'\right]^{3/2}}, \tag{1.11}$$

where parameter $-1 \leq g \leq 1$ characterizes the bias in the scattering angle, ranging from backscattering ($g \to -1$) through isotropic scattering ($g = 0$) to forward scattering ($g \to 1$). It can be easily verified that the resulting probability density function is normalized, namely

$$\frac{1}{c\,\mu_s(\mathbf{r})} \int_{4\pi} K(\hat{\mathbf{n}}, \hat{\mathbf{n}}'|\mathbf{r})\, d\Omega' = \frac{1 - g^2}{4\pi} \int_0^{2\pi} \int_0^{\pi} \frac{\sin\theta}{\left[1 + g^2 - 2g\cos\theta\right]^{3/2}}\, d\theta\, d\phi = 1$$

for all $\hat{\mathbf{n}}$ and $\mathbf{r}$.

1.1.3 *Compton scattering*

For high-energy radiation, such as X-rays or γ-rays, the prevailing form of interaction with the medium is Compton scattering. This is especially true in materials with a low atomic number, in particular biological tissue (Sethi, 2006).

In Compton scattering, an incident photon interacts with a free or loosely bound electron at rest in the outer shell of an atom of the medium. During the interaction, the photon changes its direction and transfers a portion of its energy to the electron, which recoils due to the momentum and energy acquired from the photon (Compton, 1923) (see Figure 1.2).

Using conservation of energy and momentum, the energy E of the scattered photon can be expressed through the energy E' of the incident photon and its scattering angle θ_s as follows (Krane, 2019):

$$\frac{1}{E} = \frac{1}{E'} + \frac{1}{mc^2}(1 - \cos\theta_s), \tag{1.12}$$

where m is the mass of an electron at rest. Consequently, the kernel K of the scattering term (1.8) must include a delta function supported on the

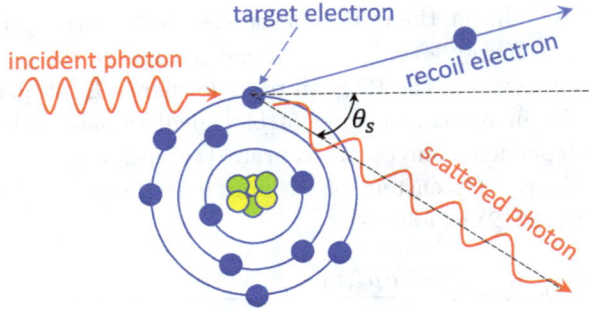

Fig. 1.2   A sketch of Compton scattering (also called Compton effect).

solution set of equation (1.12). In fact, one can show that in isotropic media,

$$K(\hat{\mathbf{n}}, E; \hat{\mathbf{n}}', E'|\mathbf{r}) = \tilde{K}(\theta_s, E, E'|\mathbf{r}) =$$

$$c\, n_s(\mathbf{r}) \frac{\partial \sigma_s}{\partial \Omega}(\theta_s, E')\, \delta\left\{ E - \left[\frac{1}{E'} + \frac{1}{mc^2}(1 - \cos\theta_s) \right]^{-1} \right\}, \quad (1.13)$$

where n_s is the *density of the scatterers* (free electrons) in the medium and $\partial \sigma_s / \partial \Omega$ is the *differential scattering cross section* (Barrett and Myers, 2003). The latter characterizes the distribution of scattering angles for photons of a given energy scattering from a single electron. It satisfies the well-known Klein–Nishina formula (Drake, 2006):

$$\frac{\partial \sigma_s}{\partial \Omega}(\theta_s, E') = \frac{1}{2}\, r_e^2\, P(\theta_s, E')^2 \left[P(\theta_s, E') + P(\theta_s, E')^{-1} - \sin^2\theta_s \right],$$

where r_e is a constant (the *classical electron radius*) and $P(\theta_s, E')$ is the ratio of the photon energy after and before the scattering, i.e.

$$P(\theta_s, E') = \left[1 + \frac{E'}{mc^2}(1 - \cos\theta_s) \right]^{-1}.$$

As a final remark, we would like to mention that the scattering coefficient μ_s can be expressed as

$$\mu_s(\mathbf{r}, E) = n_s(\mathbf{r})\, \sigma_s(E), \quad (1.14)$$

where σ_s is the (total) *scattering cross section* obtained by integrating $\partial \sigma_s / \partial \Omega$ over the full solid angle of 4π steradians:

$$\sigma_s(E) = \int_{4\pi} \frac{\partial \sigma_s}{\partial \Omega}(\theta, E)\, d\Omega.$$

1.2 Solutions of Radiative Transport Equation

In this section, we present a method for solving the RTE using the Neumann series approach. A truncated version of that series (often called Born approximation) leads to the consideration of the generalized Radon transforms, which play a central role in this book. We start with a solution of the simplest case of the RTE and gradually build up on it, adding techniques necessary to treat the general case. In all the models considered here, we assume that the medium under consideration has a finite size, so all functions appearing in the RTE (i.e. w, q, μ_a, μ_s, and μ_t) are compactly supported in the spatial variable $\mathbf{r}$.

1.2.1 *Emission and propagation*

Let us first consider a process of photon transport without any attenuation. In that case, the steady-state equation (1.10) simplifies into the following relation:

$$c\,\hat{\mathbf{n}} \cdot \nabla w(\mathbf{r}, \hat{\mathbf{n}}, E) = q(\mathbf{r}, \hat{\mathbf{n}}, E). \tag{1.15}$$

By choosing a coordinate system where one of the coordinate axes is parallel to $\hat{\mathbf{n}}$, equation (1.15) can be viewed as a trivial ordinary differential equation (ODE), which can be solved by direct integration. Namely,

$$w(\mathbf{r}, \hat{\mathbf{n}}, E) = \frac{1}{c} \int_0^\infty q(\mathbf{r} - l\hat{\mathbf{n}}, \hat{\mathbf{n}}, E)\, dl. \tag{1.16}$$

An intuitive interpretation of the above solution is the following. In the absence of scattering and absorption, the photons located at $\mathbf{r}$ and moving in the direction $\hat{\mathbf{n}}$ are generated by all sources located on the half line passing through $\mathbf{r}$ parallel to $\hat{\mathbf{n}}$. The photons generated on the other half of the same line are moving away from $\mathbf{r}$ and thus do not contribute to $w(\mathbf{r}, \hat{\mathbf{n}}, E)$.

The integral operator on the right-hand side of equation (1.16) can be expressed through the *divergent beam transform* (Natterer, 2001b):

$$\mathcal{X}f(\mathbf{r}, \hat{\mathbf{n}}) \doteq \int_0^\infty f(\mathbf{r} + l\hat{\mathbf{n}})\, dl, \tag{1.17}$$

mapping a function $f : \mathbb{R}^d \to \mathbb{R}$ to its integrals along the rays emanating from a point $\mathbf{r} \in \mathbb{R}^d$ in the direction $\hat{\mathbf{n}} \in \mathbb{S}^{d-1}$. It is assumed that the function f decays sufficiently fast at infinity for the integral in (1.17) to converge.

Many authors refer to $\mathcal{X}$ as the X-ray transform (e.g. see Barrett and Myers, 2003), but in the mathematical community, that term is typically used for the transform integrating along an entire line in $\mathbb{R}^d$, as opposed to a ray. Certainly, when applied to functions of compact support (which is always the case in medical imaging), integration along a line is equivalent to integration along a ray with an appropriately chosen vertex. However, for more general classes of functions, the distinction is indispensable. Some other names used for $\mathcal{X}$ include *fan-beam transform* (in $\mathbb{R}^2$) (Natterer, 2001b) and *cone-beam transform* (Natterer and Wübbeling, 2001).

Using the new notation, equation (1.16) can be rewritten as

$$w(\mathbf{r}, \hat{\mathbf{n}}, E) = \frac{1}{c} \left(\mathcal{X} q_{\hat{\mathbf{n}}, E} \right) (\mathbf{r}, -\hat{\mathbf{n}}), \qquad (1.18)$$

where $q_{\hat{\mathbf{n}}, E}(\mathbf{r}) = q(\mathbf{r}, \hat{\mathbf{n}}, E)$.

1.2.2 *Emission, propagation, and absorption*

Let us now add absorption to the process. In this case, the steady-state solution of the RTE satisfies the following relation:

$$\hat{\mathbf{n}} \cdot \nabla w(\mathbf{r}, \hat{\mathbf{n}}, E) + \mu_a(\mathbf{r}, E) \, w(\mathbf{r}, \hat{\mathbf{n}}, E) = \frac{1}{c} \, q(\mathbf{r}, \hat{\mathbf{n}}, E). \qquad (1.19)$$

With an appropriate change of coordinates, equation (1.19) can be treated as a linear ODE, which can be solved through multiplication by an integrating factor. Namely, let us assume (without loss of generality) that $\hat{\mathbf{n}} = (1, 0, 0)$. Then, equation (1.19) becomes

$$w' + \mu_a \, w = \frac{1}{c} \, q,$$

where $\mathbf{r} = (x, y, z)$, $w' = \partial w / \partial x$, and the remaining variables y, z, E are treated as parameters. Then, the integrating factor $I(x)$ satisfies

$$I(x) = e^{\int_{-\infty}^{x} \mu_a(\tilde{x}, \cdot) \, d\tilde{x}} \quad \text{and} \quad [wI]' = \frac{1}{c} q \, I.$$

Integrating the last expression and simplifying the result, we obtain the following:

$$w(x, \cdot) = \frac{1}{c} I^{-1}(x) \int_{-\infty}^{x} q(\tilde{x}, \cdot) \, I(\tilde{x}) \, d\tilde{x} = \frac{1}{c} \int_{-\infty}^{x} q(\tilde{x}, \cdot) \, e^{-\int_{\tilde{x}}^{x} \mu_a(l, \cdot) \, dl} \, d\tilde{x}.$$

Returning back to the original vector variables, we can write the solution of equation (1.19) as

$$w(\mathbf{r}, \hat{\mathbf{n}}, E) = \frac{1}{c} \int_0^\infty q(\mathbf{r} - l\hat{\mathbf{n}}, \hat{\mathbf{n}}, E)\, e^{-\int_0^l \mu_a(\mathbf{r} - \tilde{l}\hat{\mathbf{n}}, E)\, d\tilde{l}}\, dl. \qquad (1.20)$$

The interpretation of (1.20) is similar to that of (1.16), with the additional weight factor implying that the contribution of the photon sources to the distribution $w(\mathbf{r}, \hat{\mathbf{n}}, E)$ diminishes with distance from $\mathbf{r}$ due to absorption.

Consider a point $\mathbf{r} = \mathbf{r}_1$ on the boundary of a convex domain containing the support of q and μ_a and $\hat{\mathbf{n}}$ pointing outward (see Figure 1.3). Then,

$$q(\mathbf{r}_1 - l\hat{\mathbf{n}}, \hat{\mathbf{n}}, E) = 0, \quad \mu_a(\mathbf{r}_1 - l\hat{\mathbf{n}}, E) = 0, \quad \forall\, l < 0$$

and

$$w(\mathbf{r}_1, \hat{\mathbf{n}}, E) = \frac{1}{c} \int_{-\infty}^\infty q(\mathbf{r}_1 + l\hat{\mathbf{n}}, \hat{\mathbf{n}}, E)\, e^{-\int_l^\infty \mu_a(\mathbf{r}_1 + \tilde{l}\hat{\mathbf{n}}, E)\, d\tilde{l}}\, dl. \qquad (1.21)$$

The integral operator on the right-hand side of equation (1.21) corresponds to the *attenuated ray transform* (Natterer, 2001b):

$$\mathcal{P}_\mu f(\mathbf{x}, \hat{\mathbf{n}}) \doteq \int_{\mathbb{R}} f(\mathbf{x} + l\hat{\mathbf{n}})\, e^{-\int_l^\infty \mu(\mathbf{x} + \tilde{l}\hat{\mathbf{n}})\, d\tilde{l}}\, dl, \quad \mathbf{x} \in \hat{\mathbf{n}}^\perp, \quad \hat{\mathbf{n}} \in \mathbb{S}^{d-1}, \qquad (1.22)$$

mapping a function $f : \mathbb{R}^d \to \mathbb{R}$ to its weighted integrals along the lines passing through a point $\mathbf{x} \in \hat{\mathbf{n}}^\perp$ in the direction $\hat{\mathbf{n}} \in \mathbb{S}^{d-1}$. As usual, it is assumed that the function f decays sufficiently fast at infinity for the integral in (1.22) to converge.

Using the notation for the attenuated ray transform, we can rewrite (1.21) as

$$c\, w(\mathbf{r}_1, \hat{\mathbf{n}}, E) = (\mathcal{P}_{\mu_a} q_{\hat{\mathbf{n}}, E})(\mathbf{x}, \hat{\mathbf{n}}), \quad \mathbf{x} = \text{proj}_{\hat{\mathbf{n}}^\perp} \mathbf{r}_1. \qquad (1.23)$$

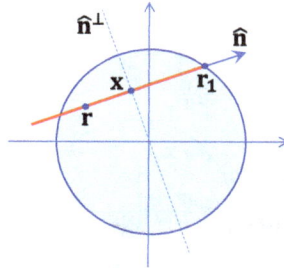

Fig. 1.3 The photon distribution in the (outward) direction $\hat{\mathbf{n}}$ at point $\mathbf{r}_1$ on the boundary of a convex image domain corresponds to the attenuated ray transform of the source q. Here, $\hat{\mathbf{n}}^\perp$ denotes the subspace normal to $\hat{\mathbf{n}}$, and $\mathbf{x}$ is the projection of $\mathbf{r}_1$ onto $\hat{\mathbf{n}}^\perp$.

1.2.3   *Propagation, out-scattering, and/or absorption*

In many applications, the medium under consideration does not have internal sources of radiation. Instead, radiation generated outside of it is sent through the medium, while a set of exterior detectors measures the intensity of the transmitted and/or scattered radiation.

A classical example is X-ray computed tomography (CT), in which a collimated emitter at point $\mathbf{r}_0$ sends a narrow beam of photons of intensity w_0 along a straight line through the patient's body. A collimated detector at point $\mathbf{r}_1$ on the opposite side of the body measures the intensity of the incoming beam (see Figure 1.4). In that case, neglecting multiply scattered particles, one can assume that the scattering term (1.6) of the RTE contains only the out-scattering component (1.7).

In the absence of internal sources and the in-scattering term, the steady-state equation (1.10) simplifies into the following:

$$\hat{\mathbf{n}} \cdot \nabla w(\mathbf{r}, \hat{\mathbf{n}}, E) + \mu_t(\mathbf{r}, E)\, w(\mathbf{r}, \hat{\mathbf{n}}, E) = 0, \tag{1.24}$$

supplemented with a boundary condition representing the external source of radiation. After an appropriate change of coordinates, equation (1.24) and its boundary condition can be treated as a separable ODE with a prescribed initial condition and solved using standard techniques.

Let us now examine the solution of the RTE in the CT setup described above. For a fixed pair of collimated emitter and detector located correspondingly at points $\mathbf{r}_0$ and $\mathbf{r}_1$, the direction of ray propagation is also fixed, namely

$$\hat{\mathbf{n}}_0 = \frac{\mathbf{r}_1 - \mathbf{r}_0}{||\mathbf{r}_1 - \mathbf{r}_0||}. \tag{1.25}$$

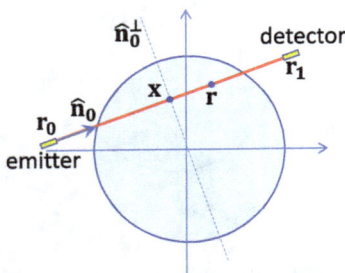

Fig. 1.4   A sketch of one collimated emitter/detector pair (out of many) used in CT.

The external source of radiation can be characterized by the condition

$$w(\mathbf{r}_0, \hat{\mathbf{n}}_0, E) = w_0(E), \tag{1.26}$$

leading to the relation

$$w(\mathbf{r}, \hat{\mathbf{n}}_0, E) = w_0(E)\, e^{-\int_{-|\mathbf{r}-\mathbf{r}_0|}^{0} \mu_t(\mathbf{r}+l\hat{\mathbf{n}}_0)dl}, \tag{1.27}$$

where $\mathbf{r}$ is any point on the line segment between $\mathbf{r}_0$ and $\mathbf{r}_1$. In particular,

$$w(\mathbf{r}_1, \hat{\mathbf{n}}_0, E) = w_0(E)\, e^{-\int_{\mathbb{R}} \mu_t(\mathbf{r}_0+l\hat{\mathbf{n}}_0)dl}, \tag{1.28}$$

where we have used the fact that μ_t is zero outside the body.

Remark 1.1. In solution (1.27), we have modified the meaning of w, treating it as a photon distribution function depending only on energy and one scalar variable defined along the line connecting the emitter and detector locations. One can certainly consider the RTE in the CT setup described above using the original definition of the photon distribution function w. However, that will require using appropriate Dirac delta functions with respect to $\hat{\mathbf{n}}$ and $\mathbf{r}$, introducing an unnecessary complication in the formula of the solution.

The integral operator on the right-hand side of equation (1.28) corresponds to the *X-ray transform* (Natterer, 2001b):

$$\mathcal{P}f(\mathbf{x}, \hat{\mathbf{n}}) \doteq \int_{\mathbb{R}} f(\mathbf{x}+l\hat{\mathbf{n}})\, dl, \quad \mathbf{x} \in \hat{\mathbf{n}}^{\perp}, \quad \hat{\mathbf{n}} \in \mathbb{S}^{d-1}, \tag{1.29}$$

mapping a function $f : \mathbb{R}^d \to \mathbb{R}$ to its integrals along the lines passing through a point $\mathbf{x} \in \hat{\mathbf{n}}^{\perp}$ in the direction $\hat{\mathbf{n}} \in \mathbb{S}^{d-1}$.

Using the notation for the X-ray transform, we can rewrite (1.28) as

$$\ln \frac{w(\mathbf{r}_0, \hat{\mathbf{n}}_0, E)}{w(\mathbf{r}_1, \hat{\mathbf{n}}_0, E)} = (\mathcal{P}\mu_t)(\mathbf{x}_0, \hat{\mathbf{n}}_0), \quad \mathbf{x}_0 = \mathrm{proj}_{\hat{\mathbf{n}}_0^{\perp}} \mathbf{r}_0. \tag{1.30}$$

Remark 1.2. It is easy to note that if the process described above does not include absorption or out-scattering, then equations (1.24) and (1.30) still hold, but μ_t should be substituted correspondingly by μ_s or μ_a.

1.2.4   *Full equation for an energy-independent setup*

Let us now consider the complete form of the RTE (1.9) in the case of monochromatic sources and elastic scattering, where all the terms of the equation are independent of energy:

$$\hat{\mathbf{n}} \cdot \nabla w + \mu_t w = \frac{1}{c}\left(q + \mathcal{K}w\right). \tag{1.31}$$

This is an integro-differential equation since it involves both derivatives and integrals of the unknown function. To find its solution, we first modify it into a purely integral form following the initial steps of the solution of (1.19).

Multiplying both sides of equation (1.31) by the exponential factor and integrating along $-\hat{\mathbf{n}}$, we get

$$w(\mathbf{r},\hat{\mathbf{n}}) = \frac{1}{c}\int_0^\infty \left[q(\mathbf{r}-l\hat{\mathbf{n}},\hat{\mathbf{n}}) + \mathcal{K}w(\mathbf{r}-l\hat{\mathbf{n}},\hat{\mathbf{n}})\right] e^{-\int_0^l \mu_t(\mathbf{r}-\tilde{l}\hat{\mathbf{n}})\,d\tilde{l}}\,dl.$$

Using notation (1.22) for the attenuated ray transform, the last expression can be written as follows:

$$w = \frac{1}{c}\mathcal{P}_{\mu_t}q + \frac{1}{c}\mathcal{P}_{\mu_t}\mathcal{K}w, \tag{1.32}$$

or equivalently

$$\left(\mathcal{I} - \frac{1}{c}\mathcal{P}_{\mu_t}\mathcal{K}\right)w = \frac{1}{c}\mathcal{P}_{\mu_t}q, \tag{1.33}$$

where $\mathcal{I}$ denotes the identity operator.

Under suitable conditions discussed in the following, equation (1.33) with a nonzero source q can be solved for w using the Neumann series:

$$w = \frac{1}{c}\left(\mathcal{I} - \frac{1}{c}\mathcal{P}_{\mu_t}\mathcal{K}\right)^{-1}\mathcal{P}_{\mu_t}q = \sum_{k=0}^\infty \left(\frac{1}{c}\mathcal{P}_{\mu_t}\mathcal{K}\right)^k \frac{1}{c}\mathcal{P}_{\mu_t}q$$

$$= \frac{1}{c}\mathcal{P}_{\mu_t}q + \frac{1}{c^2}\mathcal{P}_{\mu_t}\mathcal{K}\mathcal{P}_{\mu_t}q + \frac{1}{c^3}\mathcal{P}_{\mu_t}\mathcal{K}\mathcal{P}_{\mu_t}\mathcal{K}\mathcal{P}_{\mu_t}q + \cdots. \tag{1.34}$$

One can think of the above inversion formula as a generalization of the formula for the sum of the geometric series to the case of operators. A sufficient condition for the convergence of such an operator series is also similar to the condition for the geometric series. Namely, if $\mathcal{T}: X \to X$ is a bounded linear operator on a Banach space X and $||\mathcal{T}||$ is its norm, then $\mathcal{I} - \mathcal{T}$ is invertible and

$$(\mathcal{I} - \mathcal{T})^{-1} = \sum_{k=0}^{\infty} \mathcal{T}^k, \tag{1.35}$$

whenever $||\mathcal{T}|| < 1$. In that case, the series in (1.35) converges in the operator norm (e.g. see Cheney, 2001).

The convergence condition for series (1.34) can be verified separately in each specific setup. For example, in the case of optical imaging of biological tissue with monochromatic time-independent sources (recall (1.11)), we have:

$$\mathcal{T} = \frac{1}{c} \mathcal{P}_{\mu_t} \mathcal{K} : \mathbb{R}^3 \times \mathbb{S}^2 \to \mathbb{R}^3 \times \mathbb{S}^2$$

and

$$||\mathcal{T}||_\infty \leq \left\| \mu_s(\mathbf{r}) \int_0^\infty e^{-\int_0^l \mu_t(\mathbf{r} - \tilde{l}\hat{n})\, d\tilde{l}} dl \right\|_\infty,$$

where $|| \cdot ||_\infty$ on the right-hand side denotes the L^∞ norm of functions of $(\mathbf{r}, \hat{n})$ and $||\mathcal{T}||_\infty$ is the corresponding operator norm of $\mathcal{T}$. If we assume, in addition, that the scattering coefficient μ_s and absorption coefficient μ_a are constant, it is easy to see that

$$||\mathcal{T}||_\infty \leq \frac{\mu_s}{\mu_t} \left[1 - e^{-\mu_t R} \right] < 1,$$

where $R > 0$ is a (finite) quantity related to the size of the object. For a discussion on the convergence of the series in (1.34) using a slightly different approach and in some more general cases, we refer the reader to Florescu *et al.* (2009) and Section 2.2.2 of Duderstadt and Martin (1979).

Solution (1.34) of the full RTE (1.31) has an elucidating physical interpretation. The first term in (1.34) represents unscattered (ballistic) photons, the second term represents photons that scattered exactly once (first-order scattering), the third term those that scattered exactly twice (second-order scattering), etc. (see Figure 1.5).

A series solution to (1.31) can be obtained also for the case when $q = 0$, e.g. using the method of successive approximations applied to the integral form of that equation:

$$w = \frac{1}{c} \mathcal{P}_{\mu_t} \mathcal{K} w. \tag{1.36}$$

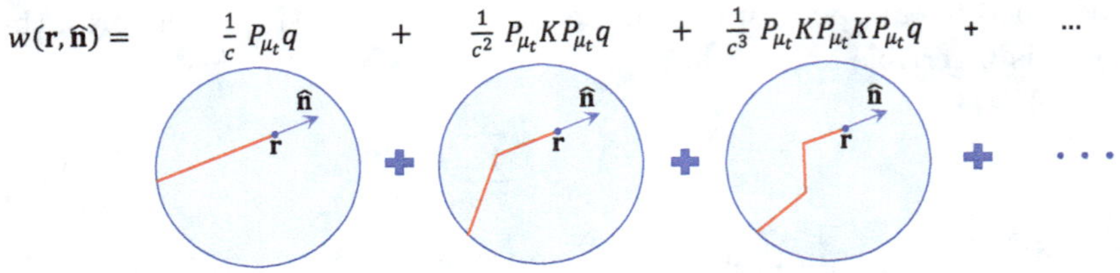

$$w(\mathbf{r}, \hat{\mathbf{n}}) = \quad \tfrac{1}{c} P_{\mu_t} q \quad + \quad \tfrac{1}{c^2} P_{\mu_t} K P_{\mu_t} q \quad + \quad \tfrac{1}{c^3} P_{\mu_t} K P_{\mu_t} K P_{\mu_t} q \quad + \quad \cdots$$

Fig. 1.5   A sketch of different orders of scattering corresponding to successive terms in the Neumann series.

Namely,

$$w = w^{(0)} + w^{(1)} + w^{(2)} + \cdots , \tag{1.37}$$

where the terms are obtained by the iterative process:

$$w^{(k+1)} = \frac{1}{c} \mathcal{P}_{\mu_t} \mathcal{K} \, w^{(k)}, \quad k = 0, 1, \dots, \tag{1.38}$$

and $w^{(0)}$ corresponds to the ballistic (unscattered) contribution to w. The requirements for convergence, as well as the physical interpretation of the terms of the series (1.37), are the same as those of (1.34).

Remark 1.3. Such expansions of scattering quantities, where the terms correspond to scattering events of different orders, are often called *Born series*. The nth partial sum of the series is called the nth *Born approximation*, or *Born approximation of order n*.

1.2.5   *Single scattering in an energy-independent case*

Let us now discuss the solution of the RTE under the assumption that all photons scatter exactly one time. By doing so, we are not assuming that the radiation process satisfies such a stringent condition but rather that the measurement of radiation is designed to account only for those photons. Such solutions are used as mathematical models of data collected in single-scattering optical tomography (SSOT) (Florescu *et al.*, 2009, 2010) and in single-scattering X-ray tomography (SSXT) (Katsevich and Krylov, 2013; Krylov and Katsevich, 2015). This section is dedicated to the model used in SSOT, where the entire process can be deemed as energy independent.

Consider a collimated emitter at point $\mathbf{r}_1$ sending a narrow beam of photons of intensity w_0 along a straight line in the direction $\hat{\mathbf{n}}_1$ into the image domain U. A detector collimated in the direction $-\hat{\mathbf{n}}_2$ measures the intensity of radiation at point $\mathbf{r}_2$ on the boundary or outside of the

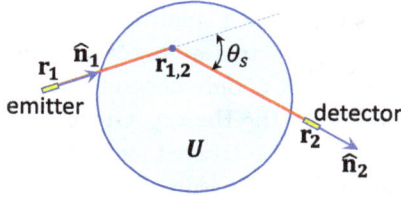

Fig. 1.6   A sketch of one collimated emitter/detector pair (out of many) used in single-scattering tomography.

image domain (see Figure 1.6). It is assumed that the "fields of view" of the emitter and detector intersect at exactly one point, i.e. there exists a unique $\mathbf{r}_{1,2} \in U$ such that

$$\mathbf{r}_{1,2} = \{\mathbf{r}_1 + l\hat{\mathbf{n}}_1, \, l > 0\} \cap \{\mathbf{r}_2 - l\hat{\mathbf{n}}_2, \, l > 0\}. \tag{1.39}$$

Disregarding photons that have scattered more than once, the data measured by the detector corresponds to the intensity $w_{1,2}$ of the single-scattered photons whose travel path is the broken ray determined by the points $\mathbf{r}_1, \mathbf{r}_{1,2}$, and $\mathbf{r}_2$. The goal of SSOT is to recover the physical characteristics of the imaged medium (e.g. μ_s and μ_t) from such single-scattered radiation corresponding to various emitter and detector locations.

To express $w_{1,2}$ in terms of μ_s and μ_t, let us recall the second term of the Born series representing the single-scattered photons:

$$w^{(1)} = \frac{1}{c} \mathcal{P}_{\mu_t} \mathcal{K} w^{(0)}. \tag{1.40}$$

The right-hand side of equation (1.40) describes three separate processes:

(1) propagation of ballistic photons of initial intensity w_0 from $\mathbf{r}_1$ to $\mathbf{r}_{1,2}$;
(2) scattering of photons at $\mathbf{r}_{1,2}$;
(3) propagation of ballistic photons from $\mathbf{r}_{1,2}$ to $\mathbf{r}_2$.

Since for a given pair of collimated emitter and receiver the travel path of the photons is fixed, we follow the idea outlined in Remark 1.1 and treat w as a one-dimensional function of photon distribution along that path. Then, by (1.27), the photon distribution at the end of the first step is characterized by the following formula:

$$w(\mathbf{r}_{1,2}, \hat{\mathbf{n}}_1) = w_0 \, e^{-(\mathcal{X}\mu_t)(\mathbf{r}_{1,2}, -\hat{\mathbf{n}}_1)}.$$

The scattering of photons at $\mathbf{r}_{1,2}$ can be expressed as follows:

$$w(\mathbf{r}_{1,2}, \hat{\mathbf{n}}_2) = \mu_s(\mathbf{r}_{1,2}) \, G(\hat{\mathbf{n}}_1, \hat{\mathbf{n}}_2) \, w(\mathbf{r}_{1,2}, \hat{\mathbf{n}}_1),$$

where $G(\hat{\mathbf{n}}_1, \hat{\mathbf{n}}_2)$ reflects the known conditional probability that a particle moving in the direction $\hat{\mathbf{n}}_1$ scatters at a given point in the direction $\hat{\mathbf{n}}_2$. Moreover, one can assume that G only depends on the scattering angle, i.e. $G(\hat{\mathbf{n}}_1, \hat{\mathbf{n}}_2) = \tilde{G}(\hat{\mathbf{n}}_1 \cdot \hat{\mathbf{n}}_2)$ (recall the Henyey–Greenstein phase function).

Finally, the third step can be treated again using formula (1.27), with the initial intensity given by the result of the scattering outlined in the second step, i.e.

$$w(\mathbf{r}_2, \hat{\mathbf{n}}_2) = w(\mathbf{r}_{1,2}, \hat{\mathbf{n}}_2)\, e^{-(\mathcal{X}\mu_t)(\mathbf{r}_{1,2},\hat{\mathbf{n}}_2)}.$$

Combining all three steps, we obtain the following relation for the measured single-scattered radiation:

$$w_{1,2} = w_0\, \mu_s(\mathbf{r}_{1,2})\, G(\hat{\mathbf{n}}_1, \hat{\mathbf{n}}_2)\, e^{-[(\mathcal{X}\mu_t)(\mathbf{r}_{1,2},-\hat{\mathbf{n}}_1)+(\mathcal{X}\mu_t)(\mathbf{r}_{1,2},\hat{\mathbf{n}}_2)]}. \qquad (1.41)$$

Remark 1.4. One can also use the original interpretation of the photon distribution function defined by equation (1.1) to obtain a relation equivalent to (1.41) using appropriate Dirac delta functions accounting for the collimation of the emitter/detector pair (see Florescu *et al.*, 2009).

The integral operator in the exponent of the right-hand side of equation (1.41) corresponds to the *V-line transform*[1] *(VLT)* (sometimes also called *broken ray transform (BRT)*) mapping a function $f : \mathbb{R}^d \to \mathbb{R}$ to a linear combination of its divergent beam transforms emanating from a common point $\mathbf{y} \in \mathbb{R}^d$ in the directions $\boldsymbol{\gamma}_1$ and $\boldsymbol{\gamma}_2$. More formally,

$$\mathcal{B}^{\mathbf{c}} f\,(\mathbf{y}, \Gamma) \doteq c_1(\mathcal{X}f)(\mathbf{y}, \boldsymbol{\gamma}_1) + c_2(\mathcal{X}f)(\mathbf{y}, \boldsymbol{\gamma}_2), \quad \mathbf{y} \in \mathbb{R}^d, \qquad (1.42)$$

where $\mathbf{c} = (c_1, c_2) \in \mathbb{R}^2$ is such that $c_1 c_2 \neq 0$, and $\Gamma = [\boldsymbol{\gamma}_1, \boldsymbol{\gamma}_2]$ is a $d \times 2$ matrix with columns $\boldsymbol{\gamma}_j \in \mathbb{S}^{d-1}$, $j = 1, 2$.

Using the above notation, we can rewrite equation (1.41) as follows:

$$\ln\left[\frac{w_{1,2}}{w_0\, G(\hat{\mathbf{n}}_1, \hat{\mathbf{n}}_2)} \right] = \ln\left[\mu_s(\mathbf{r}_{1,2}) \right] - \mathcal{B}^{\mathbf{c}}\mu_t\,(\mathbf{r}_{1,2}, \Gamma), \qquad (1.43)$$

with $\mathbf{c} = (1, 1)$ and $\Gamma = [-\hat{\mathbf{n}}_1, \hat{\mathbf{n}}_2]$.

Relation (1.43) plays a fundamental role in SSOT, which is discussed in Chapter 2.

---

[1]The name is due to the resemblance between the path of integration (the union of two rays emanating from a common vertex) and the shape of the letter V.

1.2.6 *Single scattering in an energy-dependent case*

Let us now study a model for data collected in single-scattering X-ray tomography (Katsevich and Krylov, 2013; Krylov and Katsevich, 2015). As opposed to optical imaging, where the scattering of photons is elastic, the predominant form of interaction of X-rays with biological tissue is Compton scattering, which results in a loss of energy of the scattered photons (recall Section 1.1.3). Although the treatment of the full version of the energy-dependent RTE is complicated, its single-scattering approximation requires only a small adjustment to the energy-independent case.

Consider a data acquisition setup depicted in Figure 1.6 with a monochromatic source of radiation of energy E_1. Disregarding multiply scattered particles, the data measured by this system corresponds to single-scattered photons traveling along the broken ray defined by points $\mathbf{r}_1, \mathbf{r}_{1,2}$, and $\mathbf{r}_2$. Moreover, the scattering angle θ_s is uniquely determined by the collimation directions of the emitter/detector pair. Thus, using the energy E_1 of the incident ray and the Compton scattering formula (1.12), one can also compute the energy E_2 of the X-ray after scattering.

In other words, the energy variable E in functions $\mu_s(\mathbf{r}, E)$, $\mu_a(\mathbf{r}, E)$, and $\mu_t(\mathbf{r}, E)$ accepts only two values: E_1 before and E_2 after the scattering at $\mathbf{r}_{1,2}$. Therefore, one can eliminate the dependence of those functions on E by replacing each one of them by a pair of functions depending only on the spatial variable $\mathbf{r}$ as follows:

$$\mu_{s,1}(\mathbf{r}) = \mu_s(\mathbf{r}, E_1), \quad \text{and} \quad \mu_{s,2}(\mathbf{r}) = \mu_s(\mathbf{r}, E_2), \text{ etc.}$$

Now, emulating the arguments used in the case of SSOT, one can show that the single-scattered radiation $w_{1,2}$ measured by the system here satisfies the following relation (Krylov and Katsevich, 2015):

$$w_{1,2} = w_0 \, \mu_{s,1}(\mathbf{r}_{1,2}) \, \hat{K}(\theta_s) \, e^{-[(\mathcal{X}\mu_{t,1})(\mathbf{r}_{1,2}, -\hat{\mathbf{n}}_1) + (\mathcal{X}\mu_{t,2})(\mathbf{r}_{1,2}, \hat{\mathbf{n}}_2)]}, \qquad (1.44)$$

where $\hat{K}(\theta_s)$ is a known function (recall the Klein–Nishina formula (1.13)).

It is easy to note that the main difference between formulas (1.44) and (1.41) is the change in attenuation coefficient in the second exponential. The substitution of the weight factor $G(\hat{\mathbf{n}}_1, \hat{\mathbf{n}}_2)$ (corresponding to the Henyey–Greenstein phase function) by $\hat{K}(\theta_s)$ (derived from the Klein–Nishina formula) is mathematically not important since both functions are known.

For X-rays at energy levels above 150 keV and below 1 MeV, the attenuation of photons in tissue-like materials is almost exclusively caused by Compton scattering. In this case, it is reasonable to assume that

$$\mu_t(\mathbf{r}, E) = \mu_s(\mathbf{r}, E) = C(E) \, \mu(\mathbf{r}), \qquad (1.45)$$

where the functional dependence on energy, $C(E)$, can be accurately estimated (e.g. see Sethi, 2006). Incorporating the above relations into formula (1.44), we obtain

$$\ln\left[\frac{w_{1,2}}{w_0\, C(E_1)\,\hat{K}(\theta_s)}\right] = \ln\left[\mu(\mathbf{r}_{1,2})\right] - \mathcal{B}^{\mathbf{c}}\mu\left(\mathbf{r}_{1,2}, \Gamma\right), \qquad (1.46)$$

where $\mathbf{c} = (C(E_1), C(E_2))$ and $\Gamma = [-\hat{\mathbf{n}}_1, \hat{\mathbf{n}}_2]$.

In general, the attenuation coefficient $\mu_t(\mathbf{r}, E)$ can be expressed as a finite linear combination of known basis functions that depend only on energy, while the coefficients of basis expansion depend on the medium (Alvarez and Macovski, 1976). More formally,

$$\mu_t(\mathbf{r}, E) = C_1(E)\,\mu_1(\mathbf{r}) + \cdots + C_k(E)\,\mu_k(\mathbf{r}), \qquad (1.47)$$

where $C_j(E),\ j = 1, \ldots, k$ are known functions.

Each basis function in the above expansion corresponds to a particular type of physical interaction between the X-rays and the medium. For example, for X-rays at energy levels between 30 keV and 150 keV in tissue-like materials, the dominant forms of interaction are photoelectric effect (absorption) and Compton scattering. Therefore, in that case, the attenuation coefficient can be accurately represented by the formula

$$\mu_t(\mathbf{r}, E) = C_a(E)\,\mu_a(\mathbf{r}) + C_s(E)\,\mu_s(\mathbf{r}), \qquad (1.48)$$

where the first term corresponds to absorption and the second one to scattering.

Substituting expression (1.48) for $\mu_t(\mathbf{r}, E)$ into formula (1.44), we obtain

$$\ln\left[\frac{w_{1,2}}{w_0\,\hat{K}(\theta_s)}\right] = \ln\left[\mu_{s,1}(\mathbf{r}_{1,2})\right] - (\mathcal{X}\mu_{t,1})(\mathbf{r}_{1,2}, -\hat{\mathbf{n}}_1) \qquad (1.49)$$

$$- C_a(E_2)(\mathcal{X}\mu_a)(\mathbf{r}_{1,2}, \hat{\mathbf{n}}_2) - C_s(E_2)(\mathcal{X}\mu_s)(\mathbf{r}_{1,2}, \hat{\mathbf{n}}_2).$$

Note that the quantities on the left-hand sides of (1.46) and (1.49) are either known or measured by the system. Relations (1.46) and (1.49) play an important role in various models of SSXT discussed in Chapter 2.

Chapter 2

Scattered Particle Tomography

In this chapter, we discuss five different imaging modalities that use information provided by scattered radiation to recover various internal features of an object of interest. First, we review the setups of data acquisition in these modalities and show how the measured data relates with appropriate solutions of the radiative transport equation discussed in Chapter 1. We then demonstrate that the task of image reconstruction in each case can be reduced to an inversion of an associated generalized Radon transform.

Of particular interest in this book are the transforms integrating along curves (or over surfaces) that contain "a vertex" *inside* the support of the image function. The detailed discussion of such transforms, their inversions, and other properties are presented in the subsequent chapters of the book.

In the case of transforms with a smooth path of integration, or when the "vertex" is *outside* the support of the image function, we include in this chapter an extensive overview of existing techniques and a thorough survey of the literature on the subject.

2.1 Transmission vs. Emission

Most of the radiation-based imaging techniques are divided into two groups: *transmission* tomography and *emission* tomography (see Figures 2.1 and 2.2).

In transmission tomography, the sources and detectors of radiation probing the object of interest are outside the object. Consequently, the user has control of the input energy and direction of the rays, as well as the location, possible collimation, and orientation/direction of the detectors. In the absence of internal sources of radiation, the ray intensity drops due to scattering and/or absorption before it is measured by the detectors. The image

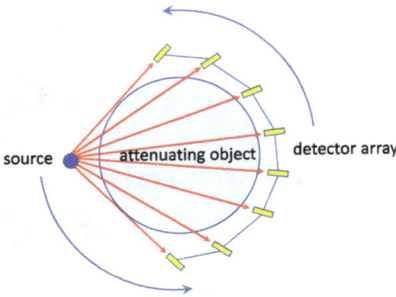

Fig. 2.1 A sketch of a transmission tomography model.

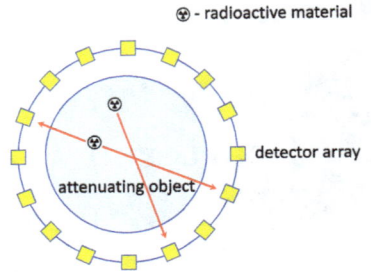

Fig. 2.2 A sketch of an emission tomography model.

reconstruction task in transmission tomography involves recovering one or several of the physical parameters of the object using radiation intensity measurements by various sets of emitters and detectors. Typically, those physical parameters are the radiation absorption coefficient μ_a, scattering coefficient μ_s, and attenuation (extinction) coefficient $\mu_t = \mu_a + \mu_s$.

A classical example of transmission tomography is X-ray computed tomography (CT), in which the measurements of the intensity of X-rays that have passed through the body are used to recover the X-ray attenuation coefficient μ_t inside the body (e.g. see Natterer and Wübbeling, 2001). Since μ_t has strong contrast between different types of tissue, a grayscale image of μ_t provides a good description of the internal structure of the body.

In emission tomography, the sources of radiation are inside the object of interest. Moreover, the source distribution function is typically the sought-after quantity, i.e. the image function. Meanwhile, the absorption coefficient μ_a and scattering coefficient μ_s are either known or are estimated through an ancillary imaging modality.

Classical examples of emission tomography used in medicine include positron emission tomography (PET) and single-photon emission computed tomography (SPECT) (e.g. see Dahlbom, 2017). In both modalities, a radiopharmaceutical emitting γ-rays is administered to a patient, typically through an intravenous injection. As the bloodstream distributes the radioactive tracer through the body, external detectors measure the intensity of radiation coming out of it. Those measurements are then used to recover the time-dependent distribution of the radiopharmaceutical in the body. The latter helps in visualizing various metabolic activities, detect internal bleeding, locate blocked arteries, etc.

The conventional models of CT, PET, and SPECT do not utilize the information from photons scattered into the field of view of the detectors, either ignoring the existence of such photons or applying various techniques to filter out their contribution to the measured data, treating it as noise (Ning *et al.*, 2004; Watson, 2000; Hutton *et al.*, 2011). However, multiple recent studies have shown that single-scattered radiation can be successfully utilized for image reconstruction in CT (Katsevich and Krylov, 2013; Krylov and Katsevich, 2015; Zhao *et al.*, 2014), PET (Berker and Schulz, 2014; Sun and Pistorius, 2014), optical imaging (Florescu *et al.*, 2009, 2010), and various other modalities (Nguyen *et al.*, 2009; Nguyen and Truong, 2010; Norton, 1994; Rigaud and Hahn, 2018; Truong and Nguyen, 2015; Webber and Lionheart, 2018).

In the subsequent sections of this chapter, we consider the mathematical models of three transmission imaging techniques (single-scattering optical tomography, single-scattering X-ray tomography, and Compton scattering tomography) and two emission imaging techniques (Compton camera imaging and Compton scattering emission tomography). While all of them use single-scattered radiation, the differences in the underlying physics and the geometry of data acquisition lead to substantially different mathematical problems required for image reconstruction. In particular, each model entails the study of a distinct generalized Radon transform.

2.2 Single-Scattering Optical Tomography

In optical tomography, a biological object under investigation is irradiated by a source of light. A portion of that light is absorbed by the body, while the rest is transmitted through and reaches the surface of the body after some scattering inside or totally unscattered. Detectors placed outside the object measure the outgoing light intensity, and that information is then used to recover the spatially varying functions of light absorption μ_a and/or scattering μ_s (Arridge and Schotland, 2009; Bal, 2009; Schotland, 2012). Due to their contrast in different tissues, these two functions provide substantial information about the internal structure of the body.

The mathematical models of optical tomography vary considerably depending on the (optical) size of the object under consideration. The size is typically described using two quantities (Ishimaru, 2017):

- *the scattering mean free path*, which is the average distance traveled by a photon between scattering events (approximately 0.1 mm in biological tissues);

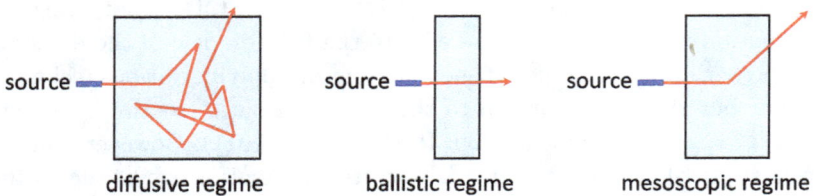

Fig. 2.3   A sketch of different regimes of light transport.

- *the transport mean free path*, which is the average distance over which the direction of propagation of the photon becomes random (approximately 1 mm in biological tissues).

An object is considered optically thick if its diameter is larger than the transport mean free path. Consequently, the vast majority of photons entering the object go through multiple scattering events (*diffusive regime*) (See Figure 2.3). In this case, the light propagation is modeled by the diffusion equation (Schotland, 2012; Van Rossum and Nieuwenhuizen, 1999).

An object is considered optically thin if its diameter is smaller than the scattering mean free path. Consequently, most photons fly through the object unscattered (*ballistic regime*) (See Figure 2.3). Here, the measured outgoing light intensity can be modeled using the X-ray transform (Schotland, 2012; Van Rossum and Nieuwenhuizen, 1999).

The intermediate case, in which the scattering mean free path is of the same order as the size of the object, is often called *mesoscopic regime* of photon transport (See Figure 2.3). Some pertinent biological applications include imaging of engineered tissues as well as small organisms, such as Zebrafish and *Drosophila melanogaster* (e.g. see Florescu *et al.*, 2009, 2010; Van Rossum and Nieuwenhuizen, 1999; Vinegoni *et al.*, 2008). In this setup, neither the X-ray transform nor the diffusion equation is applicable for modeling the light propagation. Instead, a proper approach to treat the problem requires finite-order scattering approximations of the RTE.

A novel technique of tomographic imaging in mesoscopic regime was proposed in a series of pioneering works (Florescu *et al.*, 2009, 2010, 2011). The method, called *single-scattering optical tomography (SSOT)*, uses angularly resolved measurements of single-scattered light intensity to recover the

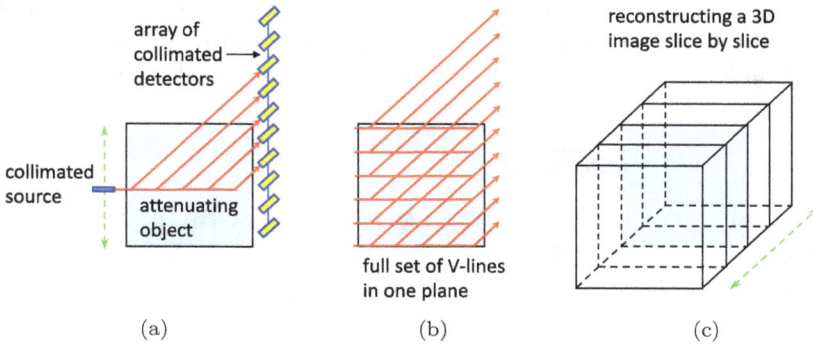

Fig. 2.4   A sketch of SSOT in a slab geometry: In (a), an array of collimated parallel detectors measures the single-scattered radiation from a collimated source sending an incident beam normal to the slab. It is assumed that after each measurement the source is shifted by a fixed distance along the vertical boundary of the slab, and the process is repeated at the new position. Figure (b) depicts a discretized subset of V-lines corresponding to a full set of such shifts. Figure (c) depicts the idea of shifting the entire system along the third dimension to generate a 3D image of the object slice by slice.

optical properties of the medium slice by slice. The main idea of the original SSOT model can be described as follows (see Figure 2.4):

- An object of investigation is illuminated with a light beam normal to its boundary.
- It is assumed that the optical thickness of the object is such that the majority of photons scatter once.[1]
- An array of collimated, parallel, co-planar detectors measures the intensity of the outgoing light on the opposite side of the object.
- The orientation of the detector array is chosen so that the field of view of each detector intersects the direction of the incident beam at a single point (scattering location) (see Figure 2.4(a)).
- A sequence of such measurements is conducted in the same plane for various locations of the source of radiation so that each point of the image domain is reasonably close to a scattering location for at least one source/detector pair (see Figure 2.4(b)).
- The measurement system (alternatively, the object of investigation) is shifted perpendicular to the data acquisition plane, and the entire process is repeated to collect data for multiple parallel slices (see Figure 2.4(c)).

---

[1]Numerical simulations have shown that the model works reasonably well for objects of thickness L up to six scattering events, i.e. $\mu_s L \leq 6$ (Florescu *et al.*, 2009).

It was shown in Section 1.2.5 that the light intensity $w_{1,2}$, measured by a given emitter/detector pair located at points $\mathbf{r}_1$ and $\mathbf{r}_2$ and collimated in the directions $\hat{\mathbf{n}}_1$ and $\hat{\mathbf{n}}_2$, satisfies the following relation (equation (1.43)):

$$\ln\left[\frac{w_{1,2}}{w_0\, G(\hat{\mathbf{n}}_1, \hat{\mathbf{n}}_2)}\right] = \ln\left[\mu_s(\mathbf{r}_{1,2})\right] - \mathcal{B}^{\mathbf{c}}\mu_t\left(\mathbf{r}_{1,2}, \Gamma\right),$$

where w_0 is the intensity of the incident beam, $G(\hat{\mathbf{n}}_1, \hat{\mathbf{n}}_2)$ is a known function, $\mathbf{r}_{1,2}$ is the scattering location (see Figure 1.6), $\Gamma = [-\hat{\mathbf{n}}_1, \hat{\mathbf{n}}_2]$, $\mathbf{c} = (1,1)$, and $\mathcal{B}^{\mathbf{c}}$ denotes the V-line transform defined by formula (1.42).

2.2.1   *A single set of detectors*

If the scattering coefficient μ_s of the medium is known or can be estimated by an ancillary imaging modality, then the SSOT measurements described above provide $\mathcal{B}^{\mathbf{c}}\mu_t\left(\mathbf{r}_{1,2}, \Gamma\right)$ at every point $\mathbf{r}_{1,2}$ of the image domain, with Γ defined by the collimation of the system. Thus, one can recover the attenuation coefficient μ_t (hence, also the absorption coefficient μ_a) of the medium by inverting the V-line transform.

Remark 2.1. In the SSOT setup described in Section 2.2, the image is reconstructed slice by slice, and all mathematical tasks can be studied in 2D. In particular, the problem of inverting the V-line transform can be considered in a given plane. Namely, one needs to recover a function $f(\mathbf{x})$ supported inside a disc $D \subset \mathbb{R}^2$ from $g(\mathbf{y}) = \mathcal{B}^{\mathbf{c}}f(\mathbf{y}, \Gamma)$, $\forall \mathbf{y} \in D$, where $\mathbf{c} = (1,1)$ and $\Gamma = [-\hat{\mathbf{n}}_1, \hat{\mathbf{n}}_2]$ is a known, constant 2×2 matrix (see Chapter 3).

Note that the task of VLT inversion formulated above is *a formally determined inverse problem*, i.e. one seeks to recover a function $f(x_1, x_2)$ of two variables from a two-dimensional (2D) data set $g(y_1, y_2)$. Such problems of inverting a generalized Radon transform from data, which has as many degrees of freedom as the dimensionality of the image function, are the centerpiece of integral geometry.

In general, the full set of V-lines in the plane is four-dimensional (4D) and so is the data corresponding to the unrestricted VLT of a function $f(\mathbf{x})$ in the plane. Indeed, a pair of variables (y_1, y_2) is needed to describe the location of the vertex of the V-line and another pair to identify the ray directions $\gamma_1, \gamma_2 \in \mathbb{S}^1$. Thus, the inversion of the full VLT in the plane is an overdetermined problem. Similarly, the set of V-lines in $\mathbb{R}^3$ is seven-dimensional, and reconstructing a three-dimensional (3D) function from its

unrestricted VLT is an overdetermined problem. To get a formally determined inverse problem in $\mathbb{R}^3$, one would need to reduce the set of V-lines
used in the (restricted) VLT to a 3D subset.

Theoretically, there are numerous ways of restricting the set of V-lines to
a subset of appropriate dimension, e.g. constraining the locations of vertices,
assigning the opening (or scattering) angles, and fixing or limiting the axes
of symmetry. In practice, the specific choice is usually made based on the
application at hand and mathematical considerations, e.g. the possibility
and level of difficulty of inverting the transform.

For example, in SSOT, it is impossible to collect full VLT data since
in order to angularly resolve the measured radiation, the detectors have to
be collimated. As a result, at each point of the boundary of the object,
photons coming only from one direction can be registered, reducing the
dimensionality of the measured data by one in $\mathbb{R}^2$ and by two in $\mathbb{R}^3$. Similar restrictions are also often applied to the direction of the incoming radiation. One such choice is to send the incident rays exclusively along the
normal to the boundary. Such setups have been proposed for SSOT in rectangular (Florescu *et al.*, 2009, 2010, 2011) and circular (Ambartsoumian,
2012; Ambartsoumian and Moon, 2013; Ambartsoumian and Roy, 2016)
geometries of data acquisition (see Figures 2.4 and 2.5).

It is natural to ask what other data acquisition geometries in 2D or
3D may be feasible for SSOT. For example, there does not seem to be
any obvious reason why all detectors have to be collimated parallel to each

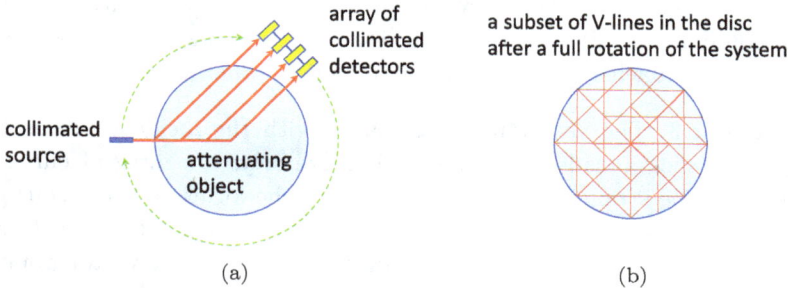

(a)   (b)

Fig. 2.5   A sketch of SSOT in a circular geometry: In (a), four collimated parallel detectors measure the single-scattered radiation from a collimated source sending an incident
beam normal to the disc. It is assumed that after each measurement, the source and
detector array are synchronously rotated by a fixed angle around the object, and the
process is repeated at the new position. Figure (b) depicts a discretized subset of V-lines
corresponding to a setup with one source and three detectors rotated around the object
with a rotation step size of 45°.

other. Are there any necessary conditions that the VLT corresponding to SSOT data must satisfy? Before answering these questions, let us first state an important observation.

Remark 2.2. It will be shown in later chapters of this book that the presence of a vertex in the paths of integration of the VLT plays a significant (and often essential) positive role, contributing to the stability of its inversion. Therefore, in image reconstruction problems of single-scattering transmission tomography, we will study only those setups where each point of the image domain is a scattering location for at least one source/detector pair.

Under the aforementioned constraint, the set of vertices of the V-lines used in the associated (restricted) VLT already has the same dimensionality as the image function. Thus, if one is interested only in data acquisition geometries that lead to a formally determined inversion of the VLT, the ray directions of all V-lines must be predetermined (no degrees of freedom are left for them). For simplicity, we assume that at each point of the image domain the ray directions of the V-lines are defined uniquely. In other words, $\gamma_1(\mathbf{y})$ and $\gamma_2(\mathbf{y})$ will be considered as known unit vector fields.

Hence, a general description of VLTs, associated with SSOT using a single set of detectors in $\mathbb{R}^d$, $d = 2, 3$, can be written as follows:

$$\mathcal{B}_\Gamma f(\mathbf{y}) \doteq \mathcal{B}^c f(\mathbf{y}, \Gamma), \tag{2.1}$$

where $\mathbf{c} = (1, 1)$ and $\Gamma(\mathbf{y}) = [\gamma_1, \gamma_2](\mathbf{y})$ is a known $d \times 2$ matrix function with columns $\gamma_j : \mathbb{R}^d \to \mathbb{S}^{d-1}$, $j = 1, 2$. To distinguish from the general case, the VLTs with $\mathbf{c} = (1, 1)$ are often called *ordinary* or *unweighted*.

Remark 2.3. The vector field associated with the incident beams (say $\gamma_1(\mathbf{y})$) satisfies an additional restriction. Namely, the integral curves of $\gamma_1(\mathbf{y})$ must be straight lines passing through the location of the source of radiation (e.g. see Figures 2.4 and 2.5). Indeed, at any point (i.e. scattering location) $\mathbf{y}_{1,2}$ inside the image domain, $\gamma_1(\mathbf{y}_{1,2})$ is a unit vector pointing toward the source of radiation located at some point $\mathbf{y}_1$. It is easy to see that for any other point $\mathbf{y}$ on the line segment between $\mathbf{y}_1$ and $\mathbf{y}_{1,2}$, $\gamma_1(\mathbf{y})$ has to point in the same direction, i.e. $\gamma_1(\mathbf{y}) = \gamma_1(\mathbf{y}_{1,2})$.

An interesting example of a data acquisition geometry satisfying the above conditions but with curved arrays of detectors was introduced by Katsevich and Krylov (2013). Here, the detectors are located either on a convex or a concave arc and are collimated along the lines with a common (focal) point (see Figure 2.8). We will review this setup in more detail in Section 2.3, covering single-scattering X-ray tomography.

Each of the systems described above has its own pros and cons, both from an engineering point of view and in terms of the mathematical problem of inverting the associated VLT. For example, the design for slab geometry does not require any movement of the detector array to image a single slice. The associated VLT is shift-invariant, which hints at the possibility of using the Fourier transform to derive an exact inversion formula for it (see Section 3.2). The circular geometry of data acquisition allows imaging of objects twice as thick as the slab system by virtue of sources and detectors having access to the object from all sides. The corresponding VLT is rotation-invariant, which leads to an inversion formula based on the Fourier series (see Section 4.1).

2.2.2 *Two different sets of detectors*

In SSOT with one set of detectors, it was necessary to know the scattering coefficient μ_s in order to find the attenuation coefficient μ_t through inversion of the corresponding VLT. In SSOT with two sets of detectors, one can recover simultaneously both μ_s and μ_t (hence also the absorption coefficient $\mu_a = \mu_t - \mu_s$).

Consider a system in slab geometry with two arrays of detectors on a line parallel to the vertical side of the slab. Each array is collimated along one fixed direction so that the field of view of any detector intersects the ray representing the incident beam at a single point. The source and detector arrays are synchronously moved along the vertical boundary of the object so that each point of the image domain ends up being a scattering location simultaneously for one detector in each array (see Figure 2.6).

Let $\hat{n}_0$ denote the direction of the incident beam, while $\hat{n}_1$ and $\hat{n}_2$ denote correspondingly the collimation directions of the first and second detector arrays. Then, by equation (1.43), the single-scattered light intensities w_{ABC} and w_{ABD}, corresponding to the common scattering location B and measured by detectors C and D in different arrays (see Figure 2.6), satisfy the following relations:

$$\ln\left[\frac{w_{ABC}}{w_0\,G(\hat{n}_0, \hat{n}_1)}\right] = \ln\left[\mu_s(B)\right] - \mathcal{B}_{\Gamma^1}\mu_t\,(B), \qquad (2.2)$$

$$\ln\left[\frac{w_{ABD}}{w_0\,G(\hat{n}_0, \hat{n}_2)}\right] = \ln\left[\mu_s(B)\right] - \mathcal{B}_{\Gamma^2}\mu_t\,(B),$$

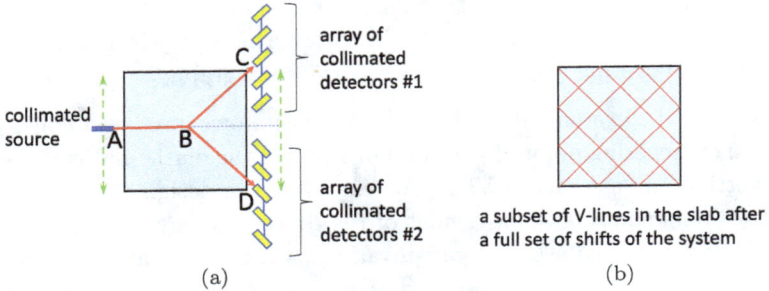

(a)

(b)

a subset of V-lines in the slab after
a full set of shifts of the system

Fig. 2.6   A sketch of SSOT with two sets of detectors in a slab geometry: In (a), five
collimated parallel detectors in each array measure the single-scattered radiation from
a collimated source sending an incident beam normal to the vertical boundary of the
slab. It is assumed that after each measurement, the source and both detector arrays
are synchronously shifted by a fixed distance along the vertical boundary of the object,
and the process is repeated at the new position. Figure (b) depicts a discretized subset
of V-lines associated with the signed VLT corresponding to a setup with one source and
three detectors in each array.

where w_0 is the intensity of the incident beam, G is a known function, $\Gamma^1 = [-\hat{\mathbf{n}}_0, \hat{\mathbf{n}}_1]$, $\Gamma^2 = [-\hat{\mathbf{n}}_0, \hat{\mathbf{n}}_2]$, and $\mathcal{B}_\Gamma$ denotes the V-line transform defined by formula (2.1).

Subtracting the first equation from the second yields

$$\ln\left[\frac{G(\hat{\mathbf{n}}_0, \hat{\mathbf{n}}_1)\, w_{ABD}}{G(\hat{\mathbf{n}}_0, \hat{\mathbf{n}}_2)\, w_{ABC}}\right] = \mathcal{B}_\Gamma^- \mu_t\,(B), \tag{2.3}$$

where $\Gamma = [\hat{\mathbf{n}}_1, \hat{\mathbf{n}}_2]$ and

$$\mathcal{B}_\Gamma^- f(\mathbf{y}) \doteq \mathcal{B}^{\mathbf{c}} f(\mathbf{y}, \Gamma), \quad \mathbf{c} = (1, -1), \tag{2.4}$$

is the so-called *signed V-line transform*.

It is clear from equation (2.3) that the SSOT measurements in the setup described above generate the values of $\mathcal{B}_\Gamma^- \mu_t\,(\mathbf{y})$ at every point $\mathbf{y}$ of the image domain. Thus, one can recover $\mu_t(\mathbf{x})$ inside the object by inverting the signed VLT (e.g. see Section 3.3.3). Once μ_t is known, one can use equation (2.2) to recover μ_s and μ_a.

Note that the restricted set of V-lines, corresponding to the integration trajectories of the aforementioned signed VLT, has the same features as the set of V-lines used in the SSOT in a slab with a single set of detectors. Namely, the V-lines have arbitrary locations of vertices but fixed directions of branches (thus a fixed opening angle and a fixed direction of the axis of symmetry). However, this is not the only option. Similar models have been studied in the case of *curved* detector arrays, leading to a different 2D family of V-lines (see Section 2.3 and Krylov and Katsevich, 2015).

Remark 2.4. As we discussed in Chapter 1, in many practical situations, $G(\hat{\mathbf{n}}_1, \hat{\mathbf{n}}_2) = \tilde{G}(\hat{\mathbf{n}}_1 \cdot \hat{\mathbf{n}}_2)$ for any unit vectors $\hat{\mathbf{n}}_1$ and $\hat{\mathbf{n}}_2$. In such cases, one can build the system so that $\hat{\mathbf{n}}_0 \cdot \hat{\mathbf{n}}_1 = \hat{\mathbf{n}}_0 \cdot \hat{\mathbf{n}}_2$, reducing equation (2.3) to

$$\ln\left[\frac{w_{ABD}}{w_{ABC}}\right] = \mathcal{B}_\Gamma^- \mu_t(B). \tag{2.5}$$

Therefore, the attenuation coefficient μ_t can be recovered without any (additional) knowledge about $\tilde{G}$.

2.2.3 *Three or more sets of detectors*

The ideas outlined in the previous section can be extended to single-scattering transmission tomography models with three or more sets of detectors (Katsevich and Krylov, 2013; Krylov and Katsevich, 2015; Zhao *et al.*, 2014). These models lead to a consideration of another generalized Radon transform, called the *star transform*, mapping a function $f : \mathbb{R}^d \to \mathbb{R}$ to a linear combination of its integrals along a finite number of rays emanating from a common vertex. More formally,

$$\mathcal{S}^{\mathbf{c}} f(\mathbf{y}, \Gamma) \doteq \sum_{k=1}^m c_k (\mathcal{X} f)(\mathbf{y}, \boldsymbol{\gamma}_k), \quad \mathbf{y} \in \mathbb{R}^d, \tag{2.6}$$

where $\mathbf{c} = (c_1, \ldots, c_m) \in \mathbb{R}^m$ is such that $\prod_{k=1}^m c_k \neq 0$ and $\Gamma = [\boldsymbol{\gamma}_1, \ldots, \boldsymbol{\gamma}_m]$ is a $d \times m$ matrix with columns $\boldsymbol{\gamma}_j \in \mathbb{S}^{d-1}$, $j = 1, \ldots, m$. The set

$$\bigcup_{k=1}^m \{\mathbf{y} + t\boldsymbol{\gamma}_k, \ t \geq 0\} \tag{2.7}$$

is called a star with a vertex at $\mathbf{y}$ and rays (or branches) along directions $\boldsymbol{\gamma}_1, \ldots, \boldsymbol{\gamma}_m$. In the case where $\boldsymbol{\gamma}_j$s are known vector fields, we use

$$\mathcal{S}_\Gamma^{\mathbf{c}} f(\mathbf{y}) \doteq \mathcal{S}^{\mathbf{c}} f(\mathbf{y}, \Gamma). \tag{2.8}$$

It is easy to note that the VLT is a particular example of the star transform, corresponding to the case of $m = 2$. It will be shown in Chapter 5 that the problem of inverting a star transform with fixed directions of rays has certain inherent pitfalls if m is even (see also Ambartsoumian and Latifi,

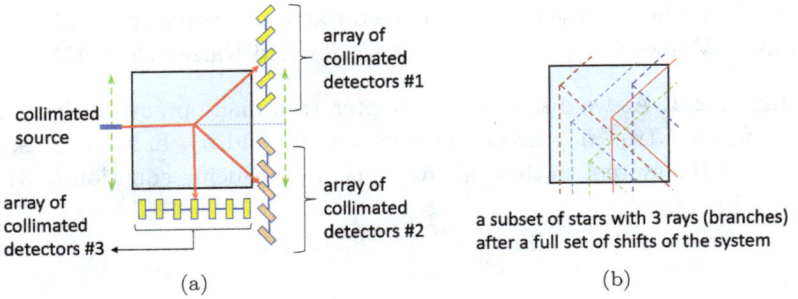

array of
collimated
detectors #1

collimated
source

array of
collimated
detectors #3

array of
collimated
detectors #2

a subset of stars with 3 rays (branches)
after a full set of shifts of the system

(a)

(b)

Fig. 2.7 A sketch of SSOT with three sets of detectors in a slab geometry: In (a), collimated parallel detectors in each array measure the single-scattered radiation from a collimated source sending an incident beam normal to the vertical boundary of the slab. It is assumed that after each measurement, the source as well as the first and second detector arrays are synchronously shifted by a fixed distance along the vertical boundary of the object, and the process is repeated at the new position. Figure (b) depicts a subset of stars associated with the star transform corresponding to the setup described in (a).

2021; Zhao *et al.*, 2014). Those drawbacks can be avoided in judiciously chosen configurations of star transforms with an odd number of rays. The latter fact provides an additional motivation to consider single-scattering tomography models with multiple detectors.

As a simple example, let us consider an SSOT setup with three arrays of collimated detectors, as depicted in Figure 2.7. It is essentially the same model that we considered in the slab geometry with two arrays of collimated detectors but with an extra array mounted at the bottom of the slab.

Adapting the notations and arguments used in the case of SSOT with two detectors, we obtain

$$\ln\left[\frac{w_{ABC_k}}{w_0\,G(\hat{\mathbf{n}}_0,\hat{\mathbf{n}}_k)}\right] = \ln\left[\mu_s(B)\right] - \mathcal{B}_{\Gamma^k}\mu_t(B), \quad k = 1,\ldots,3, \qquad (2.9)$$

where w_{ABC_k} denotes the intensity of radiation generated at point A, scattered at point B and measured by a detector at point C_k in the kth array. As before, it is assumed that each point of the image domain happens to be the scattering point B for at least one measurement.

For brevity, let us denote the known quantity corresponding to the left-hand side of equation (2.9) by

$$t_k(B) \doteq \ln\left[\frac{w_{ABC_k}}{w_0\,G(\hat{\mathbf{n}}_0,\hat{\mathbf{n}}_k)}\right].$$

Now, consider a vector $\mathbf{c} = (c_1, c_2, c_3)$ such that $\sum\limits_{k=1}^{3} c_k = 0$ and $\prod\limits_{k=1}^{3} c_k \neq 0$.

For example, $c_1 = 1$, $c_2 = -2$, and $c_3 = 1$. Then, by taking a linear combination of the available data,

$$\sum_{k=1}^{3} c_k \, t_k(\mathbf{y}) = -\mathcal{S}_\Gamma^c \, \mu_t(\mathbf{y}), \tag{2.10}$$

we generate the corresponding star transform of the attenuation coefficient μ_t. Inversion of that star transform will lead to μ_t, which then can be used in (2.9) to recover μ_s.

Remark 2.5. The ideas expressed above can be generalized to similar setups with more than three sets of detectors.

Remark 2.6. There are many other data acquisition geometries of single-scattering transmission tomography in which image reconstruction requires the inversion of some type of star transform (e.g. see Zhao *et al.*, 2014).

2.3   Single-Scattering X-ray Tomography

The conventional models of X-ray tomography are based on the assumption that the measured radiation corresponds exclusively to transmitted photons that have not interacted with matter. In other words, the photons scattering into the field of view of the detectors are neglected. The resulting mathematical problem of image reconstruction boils down to inverting the X-ray transform of the attenuation coefficient μ_t (see Section 1.2.3). However, it is well known that X-rays scatter significantly in biological tissues (Sethi, 2006), and the simplified ballistic model described above leads to imperfect sensitivity and specificity of the imaging modality (Masahiro *et al.*, 2001). Meanwhile, the scattered radiation carries valuable information about the medium, which can be used to recover interior features of the object by itself or in combination with data from transmitted radiation. Moreover, certain portions of scattered radiation can be measured with typical CT systems, for example, by utilizing off-beam idle detectors in the detection ring (Busono and Hussein, 1999).

Multiple researchers have proposed various advanced models using the scattered X-ray photons for image reconstruction (e.g. see Busono and Hussein, 1999; Cong and Wang, 2011; Katsevich and Krylov, 2013; Krylov and Katsevich, 2015; Norton, 1994; Wang *et al.*, 1999; Yuasa *et al.*, 1997; Zhao *et al.*, 2014). In this section, we discuss some of those models that are based on single-scattered radiation measured by a system with collimated emitters and detectors. We call this type of imaging modalities *single-scattering X-ray tomography (SSXT)*.

2.3.1   Low-energy X-rays

The physical nature of X-ray scattering in a medium depends largely on the energy level of the incident beam. For low X-ray energies (below 30 keV), the photon scattering is predominantly elastic (Rayleigh scattering),[2] i.e. the photons change direction, but their energy remains the same (Cong and Wang, 2011). Therefore, in SSXT performed with low-energy X-rays, one can characterize the measured data using the single-scattering approximation of the energy-independent version of the RTE (see Section 1.2.5). The resulting mathematical models are almost identical to those of SSOT discussed in the previous section. The only notable difference is that in SSXT, the incident beam does not have to be normal to the boundary of the object. In SSOT, that requirement was in place to minimize the surface reflection of the incident light, which does not happen with X-rays. As a result, the geometry of data acquisition in SSXT has more flexibility than in SSOT.

In particular, one can collect a complete data set corresponding to a 2D family of VLTs using a source located at a fixed point. While the source has to be tilted to generate incident beams in different directions, it does not have to be shifted to a new location or rotated around the body. An interesting example of such a system was described by Katsevich and Krylov (2013) (see Figure 2.8). In addition to a fixed source, the setup features two types of curvy arrays of detectors, collimated along the lines passing through a single focal point. In the case of a concave array, the focal point of detectors and the imaging object lay on the same side of the array. Such detectors are commonly used in conventional CT scanners. In the case of a convex array, the focal point of detectors and the imaging object lay on the opposite sides of the array. Such detectors correspond to a pinhole collimator of gamma cameras used in nuclear imaging.

The problem of image reconstruction in setups with curvy detectors can also be solved by inverting the corresponding VLT simply by following the arguments in Section 2.2.

---

[2]Such scattering is often called *coherent*.

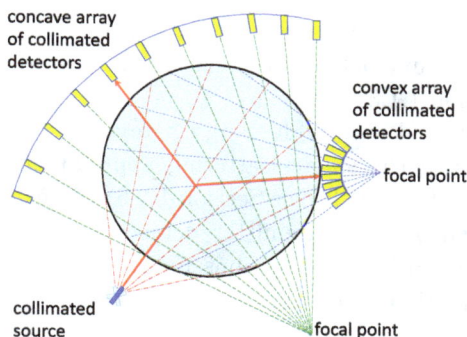

Fig. 2.8   A sketch of SSXT with convex and concave detector arrays.

In the case where the scattering coefficient μ_s of the medium is known, the measurements from a single set of detectors are sufficient for recovering the attenuation coefficient μ_t. The relevant VLT in this case is $\mathcal{B}_\Gamma f(\mathbf{y})$, where $\Gamma(\mathbf{y}) = [\gamma_1, \gamma_2](\mathbf{y})$, while $\gamma_1(\mathbf{y})$ and $\gamma_2(\mathbf{y})$ are unit vector fields directed (at each point $\mathbf{y}$ of the image domain) correspondingly toward the fixed location of the source and the focal point of the detector array. An interesting observation here, which will prove to be important later in the book, is that the integral curves of both vector fields are straight lines.

If the scattering coefficient μ_s of the medium is not known, it can be recovered simultaneously with the attenuation coefficient μ_t using measurements from two arrays of detectors. The process will require inversion of the signed VLT $\mathcal{B}_\Gamma^- f(\mathbf{y})$, where $\gamma_1(\mathbf{y})$ and $\gamma_2(\mathbf{y})$ are unit vector fields directed (at each point $\mathbf{y}$ of the image domain) correspondingly toward the focal points of the first and second arrays.

In a similar fashion, one can obtain models using three or more curvy detectors, leading to a consideration of the corresponding star transform.

Remark 2.7. Concave and convex detectors can be used in various other single or multiple array setups of SSOT and SSXT. They can also be combined with flat detector arrays, which can be thought of as having a focal point at infinity.

2.3.2   *High-energy X-rays*

For X-rays of high energy (in the range between 150 keV and 1 MeV), the predominant form of interaction with biological tissue is Compton scattering (Cooper *et al.*, 2004; Sethi, 2006), during which the photons change both direction and energy.[3] In this case, the scattering angle and the energy of the incident photons uniquely identify the (reduced) energy of the scattered photons (see Section 1.1.3).

In SSXT performed with high-energy X-rays, one can characterize the measured data using the single-scattering approximation of the energy-dependent version of the RTE (see Section 1.2.6). Namely, the X-ray intensity $w_{0,j}$, measured by a given emitter/detector pair located at points $\mathbf{r}_0$ and $\mathbf{r}_j$ and collimated in the directions $\hat{\mathbf{n}}_0$ and $\hat{\mathbf{n}}_j$, satisfies the following relation (recall equation (1.46)):

$$\ln\left[\frac{w_{0,j}}{w_0\, C(E_0)\, \hat{K}(\theta_j)}\right] = \ln\left[\mu(\mathbf{r})\right] - \mathcal{B}^{\mathbf{c}}_{\Gamma^j}\mu\,(\mathbf{r}) \doteq t_j, \qquad (2.11)$$

where w_0 and E_0 are correspondingly the intensity and energy level of the incident beam, θ_j is the scattering angle, E_j is the energy of scattered particles, $\hat{K}$ and C are known functions, $\mathbf{r}$ is the scattering location, $\mathbf{c} = (C(E_0), C(E_j))$, and $\Gamma^j = [-\hat{\mathbf{n}}_0, \hat{\mathbf{n}}_j]$. Note that all quantities on the left-hand side of (2.11) are either known or measured. Thus, the right-hand side of (2.11) can be treated as the data provided by our imaging system.

Using a setup with two arrays of detectors (e.g. see Figures 2.6 and 2.8) and subtracting their measurements corresponding to the same scattering locations, we can reduce the problem of image reconstruction to the inversion of an appropriate VLT. Namely,

$$t_1 - t_2 = \mathcal{B}^{\mathbf{c}}_{\Gamma}\mu\,(\mathbf{r}), \qquad (2.12)$$

where $\mathbf{c} = (C(E_1), -C(E_2))$ and $\Gamma = [\hat{\mathbf{n}}_1, \hat{\mathbf{n}}_2]$. Since the design of the system ensures that such measurements can be made for any $\mathbf{r}$ of the image domain, one can recover the image function μ by inverting the VLT $\mathcal{B}^{\mathbf{c}}_{\Gamma}$ (e.g. see Section 3.3.3).

Remark 2.8. If $|\theta_1(\mathbf{r})| = |\theta_2(\mathbf{r})|$ (and consequently $E_1(\mathbf{r}) = E_2(\mathbf{r})$) at each point of the image domain, then the data can be modeled using the signed VLT (2.4):

$$t_1 - t_2 = C(E_1)\,\mathcal{B}^{-}_{\Gamma}\mu\,(\mathbf{r}).$$

Remark 2.9. In the case of systems with three or more arrays of detectors (e.g. see Figure 2.7), one can use the strategy discussed in Section 2.2.3 to reduce the problem of image reconstruction to the inversion of an appropriate star transform (Zhao *et al.*, 2014).

---

[3]Such scattering is often called *incoherent*.

2.3.3 *Medium-energy X-rays*

For X-rays in the intermediate energy range (between 30 keV and 150 keV), the attenuation coefficient μ_t can be expressed as follows (recall formula (1.48)):

$$\mu_t(\mathbf{r}, E) = C_a(E)\,\mu_a(\mathbf{r}) + C_s(E)\,\mu_s(\mathbf{r}),$$

where C_a and C_s are known functions. In this case, one can use relation (1.49) to describe the system measurements. Namely, the X-ray intensity $w_{0,j}$, measured by a given emitter/detector pair located at points $\mathbf{r}_0$ and $\mathbf{r}_j$ and collimated in the directions $\hat{\mathbf{n}}_0$ and $\hat{\mathbf{n}}_j$, satisfies the following relation:

$$t_{0,j} \doteq \ln\left[\frac{w_{0,j}}{w_0\,\hat{K}(\theta_j)}\right] = \ln\left[\mu_{s,0}(\mathbf{r})\right] - (\mathcal{X}\mu_{t,0})(\mathbf{r}, -\hat{\mathbf{n}}_0)$$
$$- C_a(E_j)(\mathcal{X}\mu_a)(\mathbf{r}, \hat{\mathbf{n}}_j) - C_s(E_j)(\mathcal{X}\mu_s)(\mathbf{r}, \hat{\mathbf{n}}_j). \qquad (2.13)$$

Here, $\mu_{s,0}(\mathbf{r}) = \mu_s(\mathbf{r}, E_0)$ and $\mu_{t,0}(\mathbf{r}) = \mu_t(\mathbf{r}, E_0)$ are correspondingly the scattering and attenuation coefficients of the X-ray at the energy level E_0 of the incident beam, θ_j is the scattering angle, and E_j is the energy of scattered photons. As in all previous cases, it is assumed that the imaging system is capable of collecting such data for every point $\mathbf{r}$ of the image domain.

In general, the right-hand side of relation (2.13) cannot be reduced to an integral transform of a single function by simply using linear combinations of $t_{0,j}$ for various j (i.e. using multiple arrays of detectors). However, one can apply linear combinations of certain differential operators to the data collected from multiple arrays to recover μ_a and μ_s. The best known solution to the above problem was provided by Krylov and Katsevich (2015). It was shown therein that if the system can provide measurements from four or more arrays of detectors, the solution to the problem is considerably simplified and does not require the inversion of a generalized Radon transform. A brief outline of that method can be described as follows.

Consider an SSXT setup with $N \geq 4$ flat detector arrays and parallel incident rays (similar to the one depicted in Figure 2.7 but with more detector arrays).

Let $D_{\mathbf{n}_j} f \doteq \mathbf{n}_j \cdot \nabla f$ denote the derivative of $f : \mathbb{R}^2 \to \mathbb{R}$ in the direction $\mathbf{n}_j \in \mathbb{S}^1$, and $\xi_j \neq 0$ be some constant for all $j = 1, \ldots, N$, where $N \geq 4$. Applying $D_{\mathbf{n}_j}$ to both sides of equation (2.13) yields

$$D_{\mathbf{n}_j} t_{0,j} = D_{\mathbf{n}_j}\left\{\ln\left[\mu_{s,0}(\mathbf{r})\right] - (\mathcal{X}\mu_{t,0})(\mathbf{r}, -\hat{\mathbf{n}}_0)\right\}$$
$$+ C_a(E_j)\mu_a(\mathbf{r}) + C_s(E_j)\mu_s(\mathbf{r})$$

for any $j = 1, \ldots, N$. Taking a linear combination of the above equations with constants ξ_j produces the following relation:

$$\sum_{j=1}^{N} \xi_j D_{\mathbf{n}_j} t_{0,j} = \mu_a(\mathbf{r}) \sum_{j=1}^{N} [\xi_j \, C_a(E_j)] + \mu_s(\mathbf{r}) \sum_{j=1}^{N} [\xi_j \, C_s(E_j)]$$

$$+ \sum_{j=1}^{N} \xi_j D_{\mathbf{n}_j} \left\{ \ln \left[\mu_{s,1}(\mathbf{r}) \right] - (\mathcal{X}\mu_{t,1})(\mathbf{r}, -\hat{\mathbf{n}}_0) \right\}. \quad (2.14)$$

Since the constants ξ_j are arbitrary, one can choose various combinations of the measured data so that only one of the three terms on the right-hand side of (2.14) is nonzero, leading to an immediate recovery of μ_a and μ_s.

In particular, to get rid of the last term, one needs to choose ξ_js that satisfy the following relation:

$$\sum_{j=1}^{N} \xi_j D_{\mathbf{n}_j} = \sum_{j=1}^{N} \xi_j \mathbf{n}_j \cdot \nabla \equiv 0.$$

Using the notation $\mathbf{n}_j = (a_j, b_j)$, $j = 1, \ldots, N$, the last relation can be written as

$$\sum_{j=1}^{N} \xi_j a_j = 0, \quad \sum_{j=1}^{N} \xi_j b_j = 0. \quad (2.15)$$

To eliminate the second term of (2.14) and preserve the first, one can enforce the following restrictions:

$$\sum_{j=1}^{N} [\xi_j \, C_a(E_j)] = 0, \quad \sum_{j=1}^{N} [\xi_j \, C_s(E_j)] = 1. \quad (2.16)$$

Finally, to eliminate the first term and preserve the second, one can require

$$\sum_{j=1}^{N} [\xi_j \, C_a(E_j)] = 1, \quad \sum_{j=1}^{N} [\xi_j \, C_s(E_j)] = 0. \quad (2.17)$$

While it is theoretically possible for any of the systems (2.15), (2.16), or (2.15), (2.17) to be inconsistent, in generic situations, each of them has at least one (if $N = 4$, only one) solution set with respect to $\xi_j, j = 1, \ldots, N$. Denoting those solutions correspondingly by c_j^s and c_j^a, $j = 1, \ldots, N$, the scattering and absorption coefficients can be expressed through the measured data as follows:

$$\mu_s = \sum_{j=1}^{N} c_j^s D_{\mathbf{n}_j} t_{0,j}, \quad \mu_a = \sum_{j=1}^{N} c_j^a D_{\mathbf{n}_j} t_{0,j}. \tag{2.18}$$

We finish this section with some additional remarks about the presented method.

(1) In the setups with more than four arrays of detectors, the systems (2.15), (2.16) and (2.15), (2.17) have infinitely many solutions, producing a corresponding collection of formulas recovering μ_s and μ_a. In such cases, Krylov and Katsevich (2015) have derived an optimal choice of parameters that minimize the effects of noise in the data on the quality of reconstruction.

(2) By adding more arrays of detectors, the outlined approach can be generalized to the cases where the basis expansion (1.47) of $\mu_t(\mathbf{r}, E)$ includes more than two terms. If that expansion includes only one term (e.g. in energy-independent setups or in SSXT with high-energy X-rays), the approach works using three or more arrays of detectors.

(3) The presented method is also applicable in the case of curvy detectors with a fixed focal point and parallel geometry of incident beams.

2.4 Compton Scattering Tomography

In this section, we consider another modality of transmission tomography, called *Compton scattering tomography (CST)*, which uses radiation generated by an external monochromatic X-ray or γ-ray source and single-scattered inside a medium to recover the internal features of that medium. However, as opposed to SSXT discussed in the previous section, here the measured data is not directionally resolved since CST does not use collimated sources or detectors. Instead, the source emits a 2D fan or a 3D cone of rays, and the scattered radiation is measured by multiple energy-sensitive, omnidirectional sensors, combined into a one-dimensional (1D) array in 2D or a 2D panel in 3D (e.g. see Figure 2.9).

This modality has several advantages over previously discussed techniques. Collimated detectors used in SSXT measure only a fraction of the

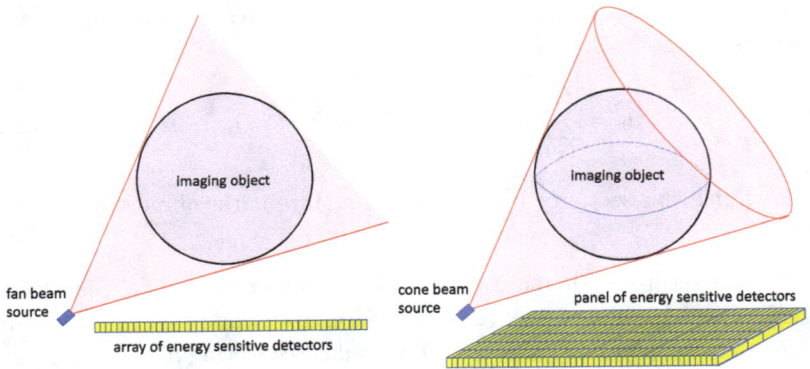

Fig. 2.9   A sketch of the basic principle of CST in 2D and 3D.

available radiation, while there is no such data loss in CST. In addition to that, the use of fan (or cone) beam source in the latter substantially speeds up the process of data acquisition. Finally, the CST setup does not require any moving parts and allows imaging the object from one side. This may be vital in various industrial and security applications where the objects of interest can be fairly large in size. However, these advantages come with a price to pay in complexity of the ensuing mathematical models.

Neglecting the photons that have scattered more than once, one can assume that the detector located at a given point $\mathbf{r}_d$ measures the energy E of single-scattered rays, which were generated by the source located at the point $\mathbf{r}_0$ and energy level E_0 (see Figure 2.10). For γ-rays and X-rays of high energy, Compton scattering is the dominant form of interaction with the medium. Thus, one can use the initial and terminal energy levels, E_0 and E, of the single-scattered rays to identity their scattering angle θ_s by Compton formula (1.12). However, the knowledge of θ_s does not uniquely determine the scattering location. Instead, it identifies a set of points, each of which can satisfy the known angular restriction for the given source–detector measurement.[4] In 2D, the locus of all such points is a circular arc, while in 3D, it coincides with an inner or an outer shell of a spindle torus, often called correspondingly a *lemon* and an *apple* (see Figure 2.10).

Consider a CST setup in $\mathbb{R}^d$ where $d = 2$ or $d = 3$. The intensity of radiation generated by a source located at $\mathbf{r}_0$, single-scattered at $\mathbf{r}_s$, and

---

[4]This set is often called a *scattering set*. In early literature, it was also called an *isogonic line* (*or curve*) (Kondic, 1978; Kondic *et al.*, 1983; Norton, 1994).

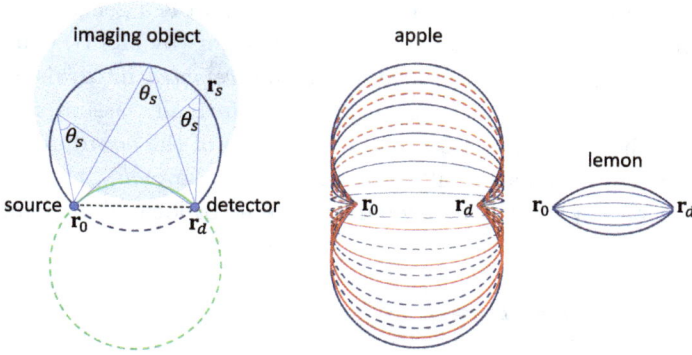

Fig. 2.10 A sketch of scattering sets in CST: The fixed points $\mathbf{r}_0$ and $\mathbf{r}_d$ represent correspondingly the locations of the source and the detector. The initial and terminal energies, E_0 and E, of the Compton scattered ray prescribe the scattering angle θ_s. The locus of points $\mathbf{r}_s$ in the plane, for which the angle $\widehat{\mathbf{r}_0\mathbf{r}_s\mathbf{r}_d}$ is θ_s, coincides with an arc of a circle uniquely identified by $\mathbf{r}_0$, $\mathbf{r}_d$, and θ_s. In the sketch above, that arc is depicted in solid blue for $\theta_s < \pi/2$ and in solid green for $\theta_s > \pi/2$. In 3D, the scattering set is a surface (a shell of spindle torus) obtained by revolving the aforementioned arc around the source–detector axis. That surface is an apple for $\theta_s < \pi/2$ and a lemon for $\theta_s > \pi/2$.

measured by a point detector located at $\mathbf{r}_d$ can be expressed as follows (recall formula (1.44)):

$$w(\mathbf{r}_0,\mathbf{r}_s,\mathbf{r}_d) = \frac{w_0\,\mu_{s,0}(\mathbf{r}_s)\,\hat{K}(\theta_s)\,e^{-[(\mathcal{X}\mu_{t,0})(\mathbf{r}_s,-\hat{\mathbf{n}}_0)+(\mathcal{X}\mu_{t,d})(\mathbf{r}_s,\hat{\mathbf{n}}_d)]}}{|\mathbf{r}_s-\mathbf{r}_0|^{d-1}\,|\mathbf{r}_d-\mathbf{r}_s|^{d-1}}, \quad (2.19)$$

where w_0 denotes (up to a known constant multiple) the intensity of the radiation source, θ_s is the scattering angle, $\hat{K}(\theta_s)$ is a known function prescribed by Klein–Nishina formula (1.13), $\mu_{s,0}(\mathbf{r}) = \mu_s(\mathbf{r}, E_0)$, $\mu_{t,0}(\mathbf{r}) = \mu_t(\mathbf{r}, E_0)$, $\mu_{t,d}(\mathbf{r}) = \mu_t(\mathbf{r}, E)$, and $\mathbf{n}_0, \mathbf{n}_d$ are unit vectors in the directions of $\mathbf{r}_s - \mathbf{r}_0$ and $\mathbf{r}_d - \mathbf{r}_s$, respectively. The denominator on the right-hand side of (2.19) is due to the *beam spreading* described by the *photometric law of distance*, also often called *inverse square law* (e.g. see Born and Wolf, 2013). It states that the intensity of radiation emitted uniformly in every direction by a point source in $\mathbb{R}^3$ is inversely proportional to the square of the distance from that source. If the source is emitting radiation uniformly in $\mathbb{R}^2$, its intensity is inversely proportional to that distance. Of course, there is no beam spreading in the models that use sources collimated along a fixed direction.

Since the detectors in CST are omnidirectional, they measure a superposition of intensities corresponding to all points $\mathbf{r}_s$ on the scattering set. As a result, the appropriate model of the measured data is a generalized Radon transform mapping the scattering coefficient μ_s, or (by (1.14)) equivalently the electron density n_s of the medium, to its *weighted* integrals along circular arcs in 2D or toric surfaces in 3D. More formally, assuming that the position of the source is fixed (e.g. at the origin of the coordinate system) and that it emits radiation at a constant energy level E_0, the number of photons at energy level E registered by a detector located at point $\mathbf{r}_d$ can be expressed as

$$I(\mathbf{r}_d, E) \doteq \int_{S(\mathbf{r}_d, E)} a(\mathbf{r}_d, E, \mathbf{r})\, n_s(\mathbf{r})\, ds, \qquad (2.20)$$

where $S(\mathbf{r}_d, E)$ is the scattering set defined by the location of the detector and the measured energy level, ds denotes the standard measure of arc length in 2D or surface area in 3D, and $a(\mathbf{r}_d, E, \mathbf{r})$ is a weight function. The latter accounts for various factors, including the dependence of differential cross section on the scattering angle $\theta_s(E)$ (see $\hat{K}(\theta_s)$ in (2.19) and formula (1.13)), beam spreading (see the denominator on the right-hand side of (2.19)), and attenuation of photons along their path from the source to the detector (see the exponential term in (2.19)). It may also reflect other parameters in more general setups, e.g. the finite size of the detector or the description of the possible anisotropy of the source (Norton, 1994; Truong and Nguyen, 2012).

If $d = 2$, the scattering sets $S(\mathbf{r}_d, E)$ correspond to circular arcs, and the integral transform defined in (2.20) is called *a (weighted) circular-arc Radon transform (CART)*. If $d = 3$, the sets $S(\mathbf{r}_d, E)$ correspond to the interior or exterior shells of spindle tori, and the integral transform defined in (2.20) is called correspondingly *a (weighted) lemon or apple transform*.

The task of image reconstruction in CST entails recovery of the spatially varying electron density of the medium $n_s(\mathbf{r})$ from the measurements of intensity $I(\mathbf{r}_d, E)$ of single-scattered radiation at multiple detector positions $\mathbf{r}_d$ and various energy levels E. In other words, one has to invert some variation of CART in $\mathbb{R}^2$ or of the apple and lemon transforms in $\mathbb{R}^3$. Typically, the detector locations are restricted to a curve or a line in the 2D case and some surface in the 3D case. With an additional degree of freedom provided by the energy variable E, the problem of inverting the corresponding generalized Radon transform is formally determined, recovering a d-dimensional function from a d-dimensional data set ($d = 2, 3$).

Currently, no exact analytical solution is known for the inverse problem corresponding to the full model, where the weight function $a(\mathbf{r}_d, E, \mathbf{r})$

accounts for the attenuation of rays along V-line trajectories. Note that it will require recovering two functions μ_s (or, equivalently, n_s) and μ_t from one generalized Radon transform. Moreover, even at energy levels at which Compton scattering is the only contributor to attenuation, inversion of the weighted CART (or weighted lemon/apple transforms) is a challenging open problem. This is not surprising since the task here is similar to the problem of inverting the attenuated Radon transform (ART) arising in SPECT. In the latter, the image function f is integrated along straight lines in $\mathbb{R}^2$ with a nonlinear weight factor that includes a second function a. While the Radon transform has been well studied with exact inversion formulas derived more than a century ago (Kuchment, 2013; Ramlau and Scherzer, 2019), ART has been extremely hard to crack. In the case of known function a, inversion formulas for ART have been derived relatively recently (Arbuzov *et al.*, 1998; Novikov, 2002a; Kunyansky, 2001; Natterer, 2001a; Boman and Strömberg, 2004), while in the case of unknown a, the problem is still largely open.

In the absence of inversion formulas for the weighted CART or weighted lemon/apple transforms, multiple exact inversion formulas have been discovered for the corresponding unweighted transforms in different geometries of data acquisition (Norton, 1994; Nguyen and Truong, 2010; Palamodov, 2011; Truong and Nguyen, 2011b, 2012; Truong, 2013; Truong and Nguyen, 2015; Webber and Lionheart, 2018; Truong and Nguyen, 2019; Webber and Miller, 2020; Tarpau and Nguyen, 2020; Tarpau *et al.*, 2020c; Cebeiro *et al.*, 2021). Simultaneously, various techniques to retrieve or compensate for beam spreading and attenuation have been suggested to resolve the discrepancy between the latter simplified models and the data measured in practice (Norton, 1994; Wang *et al.*, 1999; Rigaud *et al.*, 2012, 2013; Tarpau *et al.*, 2020b). Another approach to dealing with the full model has been the development of approximate inversion methods using microlocal analysis (Rigaud, 2017; Rigaud and Hahn, 2018; Webber and Holman, 2019; Webber *et al.*, 2020; Webber and Quinto, 2020) and optimization techniques (Webber and Quinto, 2020; Kuger and Rigaud, 2020; Rigaud and Hahn, 2021).

2.4.1 *2D models*

The first work where CST was modeled in terms of a generalized Radon transform (albeit not accounting for beam spreading and attenuation) with an exact analytical inversion was carried out by Norton (1994). It dealt with

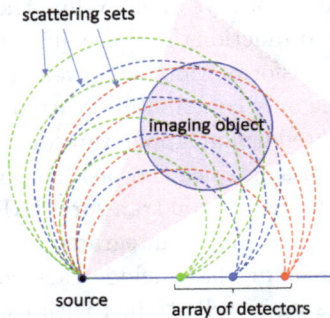

Fig. 2.11   A sketch of a 2D CST setup with a fixed source and linear array of detectors. The scattering set corresponding to a measurement made by a detector coincides with the (upper) arc of the circle, which passes through the source and detector locations and has a radius defined by the measured energy.

a 2D imaging setup, in which a fixed fan beam source illuminates a 2D slice of the imaging object. The single-scattered radiation is measured by omnidirectional, energy-sensitive detectors placed along a line that passes through the source location and does not intersect the object (see Figure 2.11).

In this setup, CART maps the image function (electron density of the object) to its integrals along circles passing through a fixed point corresponding to the radiation source.[5] An analytical inversion of that transform was derived long ago by Cormack (1963) and later generalized to the n-dimensional case by Cormack and Quinto (1980). These solutions utilized series expansions of the image function and its transform, expressing the relations between their expansion coefficients in terms of orthogonal polynomials and spherical harmonics. Norton (1994) presented an alternative inversion formula that used a filtered backprojection operator, which is more convenient for numerical implementation. The computational simulations presented in the paper not only demonstrated the efficiency of the algorithm but also highlighted the limitations of the data acquisition geometry, which has inherent instability issues akin to those in "limited-angle problem" in conventional CT (e.g. see Krishnan and Quinto, 2015). To overcome that impediment, the author advocated for systems that use more than one source. The next several setups discussed in the following use movable platforms of the sources and detectors, scanning the image from all directions.

---

[5]Since the image function is supported only on one side of the line of detectors, its integrals over circular arcs are equal to the integrals over full circles.

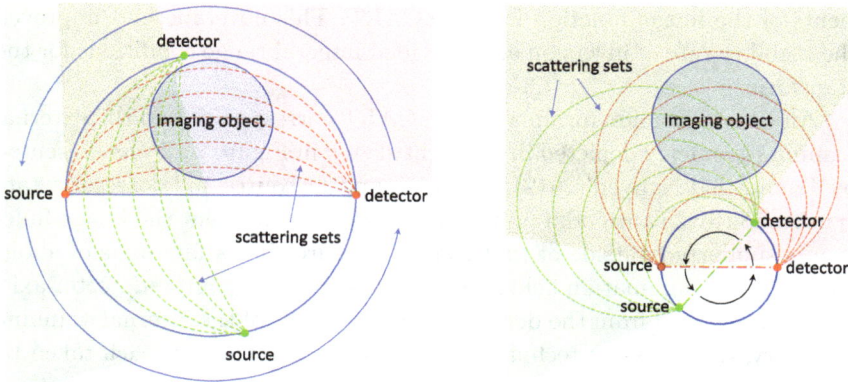

Fig. 2.12 Sketches of rotationally invariant CST models in 2D with internal scanning (on the left) and external scanning (on the right).

The first CST model based on a rotationally invariant CART and its exact analytic inversion were presented by Nguyen and Truong (2010). In this setup, a fan beam source of radiation illuminates a 2D slice of the imaging object with a fan angle of approximately 90°. The single-scattered radiation is measured by an omnidirectional detector located a fixed distance away from the source, just outside the illumination field (see the left sketch in Figure 2.12). The source and detector are synchronously rotated around an axis normal to the illumination plane and passing through the middle of the line segment connecting them. Thus, the source and detector travel as antipodal points along the scanner ring, and the measurements are repeated at each position. The system is designed so that the image function is supported inside the scanner ring.[6]

The CART in this modality maps the image function to its integrals along circular arcs located inside a given disc and subtended by antipodal points of the boundary. Nguyen and Truong (2010) provided an exact analytical inversion, recovering the Fourier coefficients of the image function from the Fourier coefficients of its CART (i.e. the measured data), in the same vein as the approach used by Cormack (1963). A different technique using equidistant curves in the hyperbolic plane was applied by Palamodov (2011) to derive an alternative formula for recovering the Fourier coeffi-

---

[6]The word "ring" here refers to the circular path of the source and detector and should not be confused with an annulus. The term is commonly used in the engineering community when referring to various imaging devices with circular scanning geometry.

cients of the image function from its CART. This novel method improved the stability rate of inversion and provided integral range conditions for the transform.

An exact inversion formula for the CART corresponding to the external scanning geometry depicted in the right sketch in Figure 2.12 was presented by Truong and Nguyen (2012). In this setup, the source and detector rotate synchronously as antipodal points on the scanner ring, but the image function is supported outside of that ring. The source has a fan angle of about 180°, and its illumination field stays on one side of the source–detector axis, slightly separated from the detector. As in the case of the internal scanning geometry, the inversion technique is very similar to the approach taken by Cormack (1963).

The success of Cormack's technique in treating the inversion of various CARTs motivated further exploration of data acquisition geometries, which can be reduced to the generalized Radon transforms discussed by Cormack (1963), Cormack (1981), and Cormack (1982). It was shown by Truong and Nguyen (2011b) that CARTs integrating along the arcs normal to the scanner ring (see Figures 2.13 and 2.14) satisfy that property and can be inverted in a similar fashion. While interesting from a theoretical point of view, the practical relevance of such setups is limited since each source–detector pair here corresponds to only one scattering angle. In other words, photons of only one energy value will be utilized by each source–detector pair, which is not efficient.

Numerical implementations of the inversion methods described above were presented by Rigaud *et al.* (2012) and Rigaud (2013).

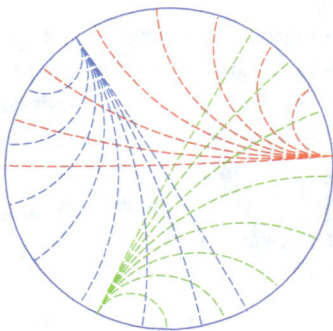

Fig. 2.13   A sketch of internal circular arcs normal to the scanner ring.

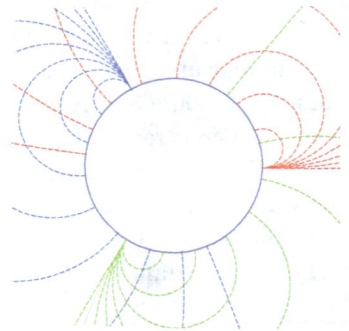

Fig. 2.14   A sketch of external circular arcs normal to the scanner ring.

In more practice-oriented research, Rigaud (2017), Tarpau *et al.* (2020a,b), and Tarpau and Nguyen (2020) considered a rotation-free modality, where CART integrates along circles passing through a fixed point. It also uses a circular geometry of data acquisition but with a single source and multiple detectors distributed around the scanner ring (see Figure 2.15). It is assumed that the detectors have the ability (e.g. by collimation) to measure the radiation coming only from one side of the source–detector axis. That allows us to eliminate the "left–right ambiguity" and record integrals along arcs corresponding to two different circles separately. Tarpau and Nguyen (2020) also considered the external scattering version of the same setup (see Figure 2.16). Rigaud (2017) presented a numerical implementation of an approximate inversion procedure for CARTs (applicable to this and previously discussed setups) recovering the image function and another approximate inversion technique recovering the contour of the image. Tarpau *et al.* (2020a,b) and Tarpau and Nguyen (2020) implemented a different reconstruction algorithm using an inversion method developed by Cormack for the generalized Radon transforms integrating along certain types of curves (Cormack, 1981, 1984), which include the circular arcs of this setup.

The essence of the method implemented by Tarpau *et al.* (2020a,b) and Tarpau and Nguyen (2020) can be tracked down to the possibility of mapping circles passing through the origin to lines by *plane inversion*

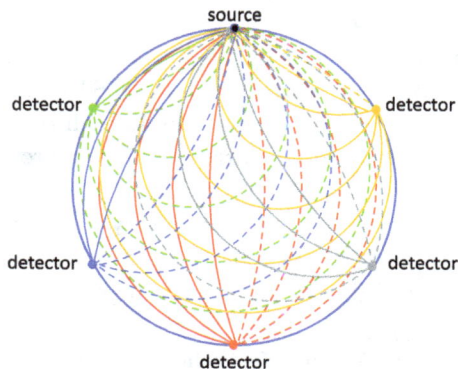

Fig. 2.15 A sketch of scattering sets in a rotation-free, internal scattering setup of CST. Solid and dashed arcs of the same color, symmetric with respect to the appropriate source–detector axis, correspond to the same scattering angle and can be a subject of "left–right ambiguity," unless the detectors have special collimation.

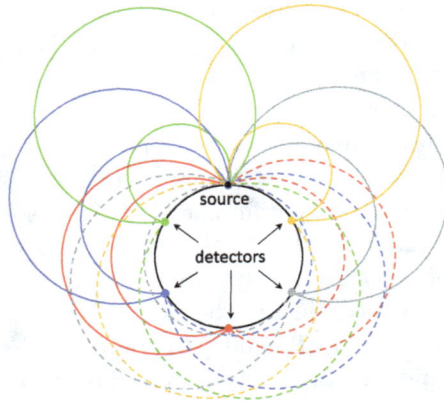

Fig. 2.16   A sketch of scattering sets in a rotation-free external scattering setup of CST. Solid and dashed arcs of the same color, symmetric with respect to the appropriate source–detector axis, correspond to the same scattering angle and can be a subject of "left–right ambiguity," unless the detectors have special collimation. It is assumed that the photons can reach the detectors by passing through the interior disc of the scanner.

diffeomorphism:

$$x \to \frac{x}{|x|^2}, \quad x \in \mathbb{R}^2 \setminus \{0\}. \tag{2.21}$$

That enables the transfer of inversion formulas and many known properties of the classical Radon transform in the plane to those for the generalized Radon transform integrating along circles passing through the same point (e.g. see Quinto, 1983; Yagle, 1992).

The same idea was also used by Cebeiro *et al.* (2017) to derive a new strategy of image reconstruction in the original CST configuration of Norton (recall Figure 2.11). In this new approach, at each detector position, one sums up the pairs of measurements corresponding to circular arcs subtended by supplementary angles. The plane inversion maps each union of such arcs into a V-line with a vertical axis of symmetry and vertices located on a line normal to that axis. The final step uses a previously developed exact inversion of a generalized Radon transform integrating functions along such V-lines (Truong and Nguyen, 2015).

A thorough study by Truong (2014) showed that all of the previously discussed CARTs (not just those corresponding to the circles passing through a fixed point) can be mapped onto the classical Radon transform with an appropriately chosen geometric transformation. For example, the internal arcs of the rotationally invariant configuration depicted in Figure 2.12 are the images of lines under the following diffeomorphism (e.g. see Webber, 2016; Truong and Nguyen, 2019)):

$$x \rightarrow \left(\sqrt{1 + \frac{1}{|x|^2}} - \frac{1}{|x|} \right) \frac{x}{|x|}, \quad x \in \mathbb{R}^2 \setminus \{0\}. \tag{2.22}$$

It is assumed here that the radius of the scanning circle is of unit length. The external arcs of the same setup can be obtained from lines using the following diffeomorphism (Truong and Nguyen, 2019):

$$x \rightarrow \left(\sqrt{1 + \frac{1}{|x|^2}} + \frac{1}{|x|} \right) \frac{x}{|x|}, \quad x \in \mathbb{R}^2 \setminus \{0\}. \tag{2.23}$$

For a geometric interpretation of these mappings, see Figure 2.17.

The relations between CARTs and the classical Radon transform have been used recently in studies of several other CST modalities. Tarpau *et al.* (2020c) propose a new model of 2D CST with a single detector rotating around a fixed, monochromatic source (see Figure 2.18). The single-scattered radiation recorded by an uncollimated detector in this setup corresponds to the sum of the integrals of electron density of the imaging object along two circular arcs passing through the source and detector locations. Two exact inversion formulas are derived for the corresponding transform on double circular arcs using techniques developed by Cormack. Similar ideas are used by Truong and Nguyen (2019) to derive the exact

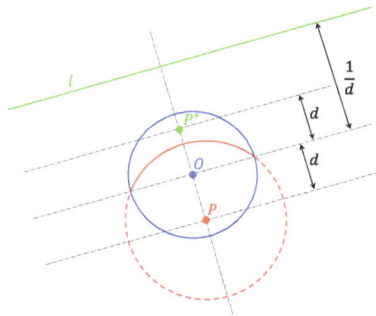

Fig. 2.17   The green line l is mapped by diffeomorphisms (2.22) and (2.23) corresponding to the internal (solid) and external (dashed) red circular arcs subtended by antipodal points of the unit (blue) circle. The blue circle centered at point O represents the scanner ring. The point P^* is symmetric to the center P of the red circle with respect to the point O. The line l is polar to the point P^* with respect to the (blue) unit circle.

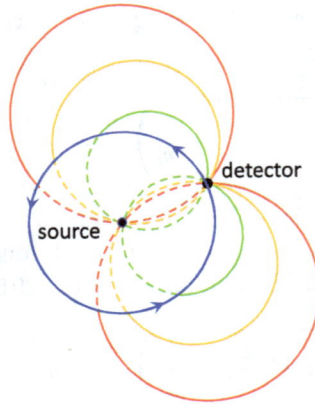

Fig. 2.18   A CST configuration with a fixed, monochromatic source and a single detector rotating around it. Since the detector is not collimated, the measurements suffer from "left–right ambiguity," and the scattering sets correspond to a union of two circular arcs.

inversion formulas for a rotationally invariant CART with radially incomplete data, recovering the image function on an annulus.

In addition to guiding the analysis of CARTs appearing in CST, connections with the classical Radon transform have been also used recently to study the inversions and properties of a generalized Radon transform integrating a 2D function over discs (Webber, 2016) and a spindle (torus) transform in $\mathbb{R}^3$ (Webber and Holman, 2019). The latter work is discussed in the next section, so let us briefly review the former.

All CST modalities that we considered earlier used monochromatic sources of radiation with a known energy level E'. As a result, the energy level E of single-scattered photons measured by a detector uniquely identified the scattering angle $\theta_s(E)$ through the Compton scattering relation (1.12). That angle, in turn, determined the circular arc of potential scattering locations of the photons, leading to the necessity of inverting the corresponding CART. Webber (2016) considered a scanning device that uses a polychromatic source of radiation, where the energy level E' of the source satisfies an inequality $E' \leq E_{\max}$, with a known upper bound $E_{\max}$. Here, instead of identifying the scattering angle θ_s, the measured energy E of single-scattered photons and Compton scattering relation (1.12) provide an upper limit $\theta_{\max}(E)$ for that angle. As a result, the set of potential scattering locations of the photons corresponds to a 2D domain enclosed by a *toric section*, i.e. it is an intersection of two discs of equal radius

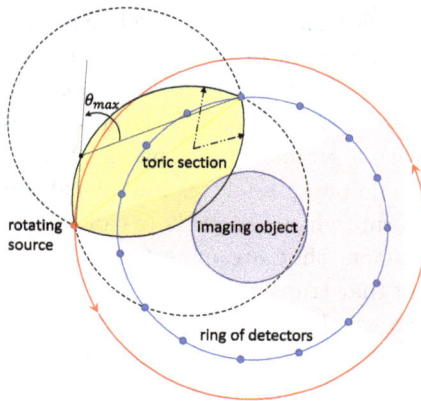

Fig. 2.19   A CST configuration with a polychromatic source rotating around the imaging object and a ring of detectors. The data corresponding to a given source–detector pair and a specific measured energy level is modeled as an integral of the electron density of the object over a 2D domain enclosed by a toric section (i.e. over an intersection of two discs of equal radius).

(see Figure 2.19). Consequently, the problem of image reconstruction in this modality requires inversion of a generalized Radon transform mapping the image function to its integrals over sets enclosed by toric sections. Due to the geometry of data acquisition and the support of the image function, this transform is equivalent to a generalized Radon transform integrating over a family of discs with a common boundary point, corresponding to a source location. The paper presents an exact inversion formula for the latter transform on a set of functions supported in an annulus centered at the source location. The result implies that data collected from a single viewpoint (i.e. from one specific position of the source) is sufficient for image reconstruction. However, as one may expect, using data from multiple views improves the quality of reconstructions. The author examines the applicability of the proposed modality to the task of baggage screening at airports and demonstrates the efficiency of the method with numerical simulations. Similar to the other models discussed in this section, the derivations here depend on a set of simplifying assumptions about the weights of the integrals associated with the measured data. The paper includes an explanation of how to reduce the artifacts caused by those simplifications.

A scanning device similar to the one depicted in Figure 2.19 was considered also by Webber and Quinto (2020). However, here the source is monochromatic, and the single-scattered radiation is measured only for those source–detector positions at which the line segment connecting them passes through the middle point of the detector ring. Therefore, the intensity measurements at a given energy level correspond to integrals over toric sections (unions of two internal circular arcs subtended by the source and detector positions) with a symmetry axis passing through the middle point of the detector ring. The paper analyzes microlocal properties of the associated integral transform (which the authors call a *toric section transform*) and explains the artifacts that are expected in any backprojection-type inversion algorithm of that transform. It also includes a proof of injectivity of the toric transform on the class of L^∞ functions supported inside the unit disc using the ideas of Cormack discussed earlier.

Another interesting design of a CST scanner, with potential application to airport baggage screening and threat detection, was considered by Webber and Miller (2020) and Webber *et al.* (2020). It uses a linear array of monochromatic sources installed above a linear array of energy-sensitive detectors to image slice by slice an object of interest placed below both the arrays. To generate a 3D image, the object is moving on a conveyor belt normal to the plane of the source and detector arrays (see Figure 2.20). The single-scattered radiation is measured only for source–detector pairs positioned along a vertical axis normal to the arrays. Consequently, the

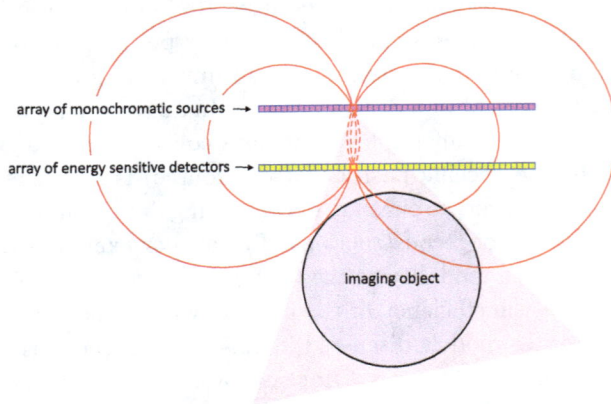

Fig. 2.20   A CST configuration with a linear array of monochromatic sources and a linear array of energy-sensitive detectors. The data corresponding to a given source–detector pair and a specific measured energy level is modeled as an integral of the electron density of the object along a toric section (i.e. a union of two (external) circular arcs of equal radius).

intensity measurements at a given energy level correspond to integrals of the object's electron density over toric sections (unions of two external circular arcs of equal radius subtended by the source and detector positions) with a symmetry axis normal to both arrays. Webber and Miller (2020) also propose an intuitive generalization of the proposed configuration to a fully 3D setup, where the linear arrays of sources and detectors are replaced by 2D arrays installed parallel to each other above the imaging object. In this case, the measured data is represented by a generalized Radon transform of the electron density of the object integrating over apple surfaces (see Figure 2.21) with a vertical axis of symmetry. Webber and Miller (2020) prove the injectivity of both 2D and 3D integral transforms mentioned above on the class of L^2 functions with compact support and derive explicit inversion formulas for them. The inversion technique is based on the Paley–Wiener–Schwartz theory of analytic continuation and an explicit inversion of 1D Volterra operators in the Fourier domain. The injectivity results are generalized to the case of the surface of revolution of a certain class of C^1 curves using the ideas of Cormack (1963). In particular, that generalization covers the case of a Radon-type transform integrating a function over conical surfaces, which appears in the models of 3D Compton scatter *emission* imaging (Truong *et al.*, 2007). The microlocal properties of the toric section transform appearing in the 2D setup described above were analyzed by Webber *et al.* (2020). Those properties were then used to derive a new, λ-tomography-type technique for an approximate reconstruction of both the electron density and attenuation coefficient of the medium using jointly the measurements of unscattered X-ray CT and Compton backscattered data.

Fig. 2.21 A spindle torus obtained by rotating a union of two intersecting circles of equal radius around the axis passing through their intersection points. The spindle torus consists of an exterior shell called an apple surface and an interior shell called a lemon surface.

We finish this section by mentioning another recent work by Kuger and Rigaud (2020), in which 2D image reconstruction is studied using CT and CST spectral data simultaneously in a scanning configuration similar to the one depicted in Figure 2.19. The proposed approach employs a combination of techniques from optimization and microlocal analysis to process the data from unscattered radiation, as well as the measured intensities of first- and second-order Compton scattering. The efficiency of the suggested method is validated via numerical implementations on synthetically generated data.

2.4.2   *3D models*

During the past couple of years, several researchers have undertaken studies on generalizing the 2D models of CST to fully 3D setups. In the latter, the 3D image of the object is recovered from a 3D transform representing the measurements of scattered radiation at once rather than slice by slice. Some of those works have already been described in the previous section (e.g. Webber and Miller, 2020), so here we briefly summarize the results of the studies not mentioned earlier.

Webber and Lionheart (2018) generalized to 3D the setups from Nguyen and Truong (2010) and Truong and Nguyen (2012) depicted in Figure 2.12. In the proposed new configuration, a monochromatic source and a detector synchronously rotate as antipodal points around the unit sphere. However, in this case, the source illuminates the entire imaging object (not just half of it, as it was in 2D). If the imaging object is located inside the sphere, the measured data is modeled as a generalized Radon transform of its electron density over lemon surfaces,[7] the tips of which rotate synchronously as antipodal points on the unit sphere (see Figure 2.22). If the imaging object is supported outside of the unit sphere, then the measured data is modeled as the corresponding apple transform of the electron density of the object. Note that in both cases, the data set has three degrees of freedom: two spherical coordinates defining the location of the source and the scattering angle identified by the energy level of the scattered radiation. Since the unknown image function is also 3D, the problem of inverting the corresponding integral transform is formally determined. The authors prove the injectivity of both transforms on a class of smooth functions supported inside an intersection of a spherical shell and a half-space. The restriction to a half-space is easy to justify since, due to the symmetry of the data acquisition geometry, every odd function is in the kernel of

---

[7]Webber and Lionheart (2018) called that transform a *spindle transform*, while Rigaud and Hahn (2018) called it a *lemon transform*.

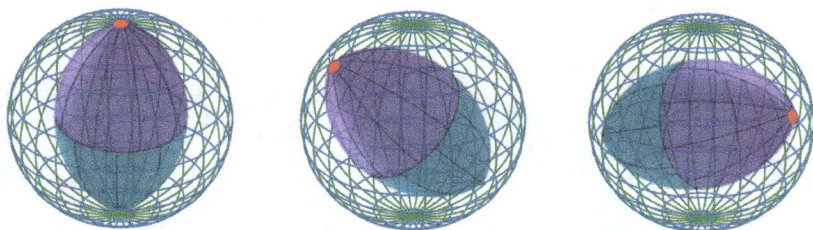

Fig. 2.22 A 3D CST configuration, where the source (marked by a red circle) and the detector synchronously rotate as antipodal points around the unit sphere. The scattering sets correspond to lemon surfaces of different radii, with tips at the source and detector positions.

the apple/lemon transforms. The proof uses an explicit inversion of a class of Volterra integral operators, relating harmonic coefficients of the image function and the measured data. The model and the injectivity results are also generalized to the case of an imaging system using a polychromatic source in the same geometry. Here, the measurements correspond to a generalized Radon transforms integrating the image function over 3D domains bounded by the lemon and apple surfaces (recall our discussion of Webber (2016) dedicated to 2D CST with polychromatic sources).

Webber and Holman (2019) studied the stability of the apple and lemon transforms introduced by Webber and Lionheart (2018) by analyzing the microlocal properties of those transforms and their normal operators (i.e. the composition of the operator and its backprojection). The work describes the artifacts in backprojection-type reconstructions from data corresponding to those transforms and provides a filter to reduce the strengths of those artifacts in reconstructed images.

Rigaud and Hahn (2018) considered multiple configurations of 3D CST and studied the properties of the associated apple and lemon transforms. The analyzed geometries of data acquisition included the rotationally invariant setup discussed above (see Figure 2.22) and three others with a fixed source and 2D arrays of detectors installed on the surfaces of a sphere (see Figure 2.23), a cylinder (see Figure 2.24), and a pair of planes (see Figure 2.25). Although the scattering sets in all presented figures here are lemon surfaces (i.e. they correspond to a scattering angle less than $\pi/2$), it is clear that if the energy of the measured radiation matches with a scattering angle larger than $\pi/2$, those sets will be represented by the corresponding apple surfaces. Rigaud and Hahn (2018) studied a class of such operators and developed a filtered backprojection-type approximate inversion technique, which preserves the contours (i.e. singularities) of the

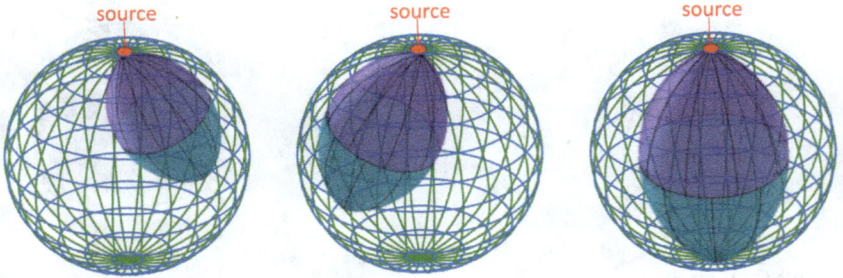

Fig. 2.23 A 3D CST configuration where the source (marked by a red circle) is fixed and a 2D array of detectors is installed on the surface of a sphere. The scattering sets correspond to lemon surfaces of various radii, with tips at the source and detector positions. This configuration is a generalization of the 2D setup considered by Rigaud (2017), Tarpau *et al.* (2020a,b), and Tarpau and Nguyen (2020) (recall Figure 2.15).

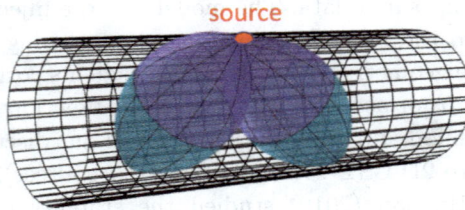

Fig. 2.24 A 3D CST configuration where the source (marked by a red circle) is fixed and a 2D array of detectors is installed on the surface of a cylinder. The scattering sets correspond to lemon surfaces of various radii, with tips at the source and detector positions.

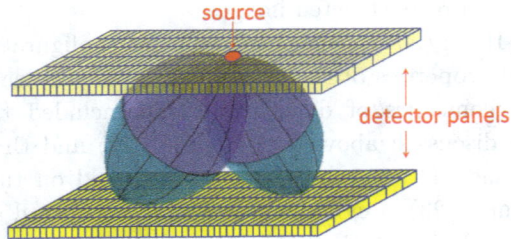

Fig. 2.25 A 3D CST configuration where the source (marked by a red circle) is fixed and a 2D array of detectors is installed on a pair of planes. The scattering sets correspond to lemon surfaces of various radii, with tips at the source and detector positions.

images function. The efficiency of the proposed approach was demonstrated by numerical simulations.

Cebeiro *et al.* (2021) studied a generalization of the 2D rotation-free external imaging modality considered by Tarpau and Nguyen (2020), where the detectors were distributed along the scanner ring represented by the unit circle, a fixed monochromatic source was located on the same ring, and the imaging object was located outside the ring (recall Figure 2.16). In the 3D configuration, the unit circle is replaced by the unit sphere, and the measured single-scattered radiation is modeled by the apple transform of the electron density of the object. The apple surfaces of integration are the complements of the lemon surfaces depicted in Figure 2.23 as parts of the corresponding spindle tori. The authors show the injectivity of that transform on the class of smooth, compactly supported function and derive an inversion formula for it. The proof uses spherical harmonic expansions and exact solutions to Abel-type integral equations relating the expansion coefficients of the image function with its transform. The results are validated by numerical experiments.

Rigaud and Hahn (2021) investigated limited data problems in CST configurations outlined in Figures 2.23 and 2.24. The first problem is related to the nonlinear dependence of the distance between different toric surfaces (corresponding to the same source and receiver pair) on the discretization step size of the scattering angle (equivalently, the measured energy). Due to a rapidly increasing derivative of the aforementioned nonlinear function, a scanner using detectors with a finite energy resolution records data that is substantially undersampled around large scattering angles (correspondingly, in the low-energy levels). The second problem, relevant only for the configuration shown in Figure 2.24, is the limited-angle issue in the data, due to the finite length of the cylindrical scanner. The paper examines the effects of data limitations on the quality of image reconstruction using filtered backprojection-type techniques and the resulting artifacts. To overcome the latter, the authors propose an alternative iterative algorithm treating the image reconstruction as a (discretized) optimization problem. They demonstrate the efficiency of that approach through numerical simulations.

Rigaud (2021) studied the possibility of using the first- and second-order scattered radiation in 3D CST with a monochromatic source to recover the contours (singularities) of the electron density of an imaging object. The measurements in the proposed model are approximated by Fourier integral operators, and the paper examines their smoothness properties. The findings of the work are validated by numerical simulations.

2.5   Compton Camera Imaging

In SSOT and SSXT discussed in Sections 2.2 and 2.3, respectively, the data was measured by physically collimated receivers. The benefit of that, of course, is the knowledge of the direction from which the photons arrive at the detector. But on the flip side, such collimation results in an extensive data loss since particles arriving at the detector from all but one direction are discarded. In CST, as discussed in Section 2.4, the flight direction of the measured photons cannot be identified since the utilized detectors are omnidirectional (apart from collimation to a plane in 2D modalities). But on the positive side, the information carried by any photon reaching the detector is taken into account. In this section, we discuss the imaging modalities that use detectors with properties fitting somewhere in between the previous two.

Compton cameras, sometimes (especially in the early literature) also called *electronically collimated cameras*, are energy-sensitive detectors which can register photons arriving from any direction. Moreover, these cameras are capable of determining the flight direction of the arriving photons up to a reflection across a fixed axis in 2D or a rotation around a fixed axis in 3D.

The simplest forms of such imaging devices have been proposed as early as in the 1970s (Schönfelder *et al.*, 1973; Todd *et al.*, 1974; Herzo *et al.*, 1975; Everett *et al.*, 1977) and 1980s (Singh, 1983; Singh and Doria, 1983). A detailed review of the early work on this subject was conducted by Phillips (1995). With the development and advancement of high-sensitivity digital detectors, Compton cameras rapidly became one of the primary tools for detection, identification, and imaging of sources of γ-radiation. They have a wide range of applications, including astrophysics (see Bandstra *et al.*, 2011, and the references therein), nuclear medicine (see Rogers *et al.*, 2004, and the references therein), and homeland security (nuclear threat detection) (Phillips, 1997; Lee and Lee, 2010; Allmaras *et al.*, 2013; Olson *et al.*, 2016). Note that all of these are emission modalities, in which the γ-rays are generated inside the medium of interest and not by an external, controlled source probing the imaging object.

A typical Compton camera consists of two parallel, digital sensor arrays, called a scatterer and an absorber (see Figure 2.26). When a photon in a γ-ray hits the scatterer at a point X', it changes its flight trajectory and hits the absorber at another point X. The sensors of the Compton camera

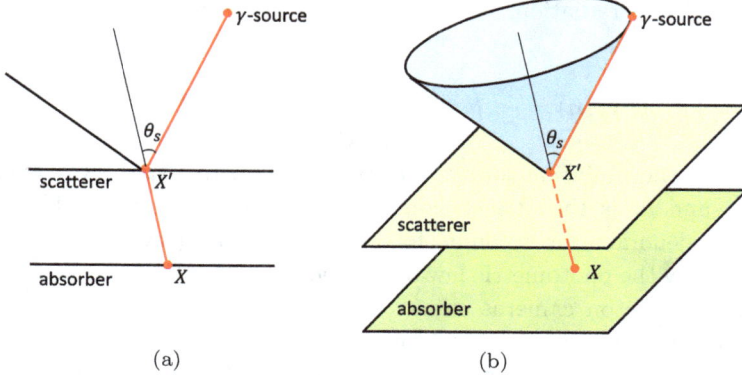

Fig. 2.26 Sketches of 2D and 3D Compton camera models: In (a), the 2D camera comprises 1D arrays of digital sensors. In (b), the 3D camera comprises 2D digital sensor plates. In general, the sensor arrays do not have to be flat and may be shaped as curves or surfaces.

register these locations X' and X as well as the energy of the particle at each point. The well-known Compton scattering relation (recall formula (1.12)) then allows to recover the scattering angle θ_s:

$$\cos\theta_s = 1 - \frac{mc^2(E' - E)}{EE'},$$

where E' is the original energy of the photon measured at X', E is the energy of the scattered photon measured at X, and m is the electron mass at rest. Filtering the data provided by the Compton camera so that it corresponds only to ballistic photons (i.e. photons that did not scatter between the source and the camera),[8] one can infer that in the 2D setup, the source of the scattered photon is located on the V-line of opening angle $2\theta_s$, with its vertex at the point X' and symmetry axis XX'. In a 3D Compton camera, one has to replace the V-line with a circular cone, i.e. the surface obtained by revolving the V-line around its axis of symmetry.

Using the version of radiative transport equation that takes into account emission, propagation, and attenuation (recall Section 1.2.2), one can model

---

[8] For example, in nuclear medicine, the initial energy of the photon is determined by the particular radiopharmaceutical used for imaging. Thus, one can filter out the scattered photons generated by the source of interest as well as those due to possible background radiation, simply based on the energy E' measured at X'.

the intensity of radiation arriving along *a single direction* $\hat{\mathbf{n}}$ to a point $\mathbf{r}_s$ of the scatterer as

$$w(\mathbf{r}_s, \hat{\mathbf{n}}) = \frac{1}{c} \int_0^\infty \frac{q(\mathbf{r}_s - l\hat{\mathbf{n}})}{l^2} \, e^{-\int_0^l \mu_t(\mathbf{r}_s - \tilde{l}\hat{\mathbf{n}}) \, d\tilde{l}} \, dl, \qquad (2.24)$$

where q is the radiation source density function supported away from the camera and μ_t is the attenuation coefficient of the medium. The factor l^2 in the denominator accounts for beam spreading (inverse square law), described by the photometric law of distance (e.g. see Born and Wolf, 2013).

Since Compton cameras cannot resolve the measurements to a single direction, the appropriate model for the intensity of radiation measured by a sensor at a given point $\mathbf{r}_s$ is the average of $w(\mathbf{r}_s, \hat{\mathbf{n}})$ over all angularly relevant directions $\hat{\mathbf{n}}$. Therefore, the task of image reconstruction in *Compton camera imaging (CCI)* requires the inversion of an *attenuated conical Radon transform*:

$$\mathcal{C}_a q \, (\mathbf{r}_s, \hat{\mathbf{s}}, \theta_s) \doteq \int_{C(\mathbf{r}_s, \hat{\mathbf{s}}, \theta_s)} a(\mathbf{r}_s, \hat{\mathbf{s}}, \theta_s, \mathbf{r}) \, q(\mathbf{r}) ds, \qquad (2.25)$$

where $C(\mathbf{r}_s, \hat{\mathbf{s}}, \theta_s)$ denotes the cone[9] with its vertex at $\mathbf{r}_s$, axis of symmetry along $\hat{\mathbf{s}} \in \mathbb{S}^{d-1}$, $d = 2, 3$, and half-opening angle $\theta_s \in (0, \pi/2)$. The integral is computed with respect to the standard measure of arc length in 2D or surface area in 3D. The weight function $a(\mathbf{r}_s, \hat{\mathbf{s}}, \theta_s, \mathbf{r})$ incorporates various factors, including the beam spreading, beam attenuation, Klein–Nishina distribution of scattering angles, and averaging constants.

Remark 2.10. In addition to photon beam spreading and attenuation, there exist several other physical phenomena that may influence the performance of a Compton camera. Some examples of such effects include Doppler broadening (Ordonez *et al.*, 1997b) and backscattering from the absorber (second plane detector array) into the scatterer (first plane detector array) (Earnhart, 1999), which were largely ignored in the early works on the subject (Chelikani *et al.*, 2004). The mathematical models of CCI have to be enhanced to incorporate these and other similar phenomena for a more accurate representation of measured data. Adding new weights to the integral geometric models will most probably make the corresponding transforms prohibitively difficult to invert. Therefore, most of the efforts on this front have been dedicated to mitigating the aforementioned negative

---

[9]If $d = 2$, the cone is substituted by the corresponding V-line. The resulting transform is called an *attenuated V-line transform*.

effects by preprocessing the measured data before applying integral geometric techniques of image reconstruction (e.g. see Smith, 2005; Frandes *et al.*, 2016).

Remark 2.11. When $\mathbf{r}_s$ is restricted to a surface or a curve located outside of supp q, the conical Radon transform (CRT) is often called *Compton transform.*

Similar to the case of CST and CART discussed in the previous section, currently, no exact analytical solution is known for the inverse problem corresponding to the full integral geometric model of CCI. Multiple researchers have studied other approaches to image reconstruction in CCI, including iterative algebraic techniques, statistical methods, and various approximate backprojection-type algorithms (e.g. see Singh and Doria, 1983; Brechner and Singh, 1990; Hebert *et al.*, 1990; Rohe *et al.*, 1997; Wilderman *et al.*, 1998; Sauve *et al.*, 1999; Wilderman *et al.*, 2001; Du *et al.*, 2001; Pauli *et al.*, 2002; Andreyev *et al.*, 2009; Frandes *et al.*, 2016). In the absence of inversion formulas for the attenuated Compton or V-line transforms, at the early stages of research on the topic, numerous exact inversion formulas have been discovered for the corresponding transforms with simpler (or without any) weights in different geometries of data acquisition. Simultaneously, various techniques to mitigate the effects of beam spreading, attenuation, and other factors built into the weight of the integral in (2.25) have been suggested to resolve the discrepancy between the latter simplified models and the data measured in practice.

2.5.1 *2D models*

The first work on 2D CCI,[10] using an integral geometric model with an exact analytic inversion of the underlying generalized Radon transform, was carried out by Basko *et al.* (1997a). The authors considered a setup with a pair of 1D, linear, parallel arrays of sensors as a scatterer and an absorber (see Figure 2.26(a)). The measurements of the camera are modeled as a 3D (thus overdetermined), weighted VLT of a function representing the radiation source distribution. The vertices of the V-lines are restricted to the line representing the scatterer, while the directions of their axes of symmetry $\hat{\mathbf{s}} \in \mathbb{S}^1$ as well as their half-opening angles $\theta_s \in (0, \pi/2)$ can be arbitrary. Assuming that beam spreading in 2D CCI results in a factor of

---

[10]Exact analytic inversions of transforms integrating along circular cones relevant for 3D CCI have been discovered earlier, e.g. Cree and Bones (1994).

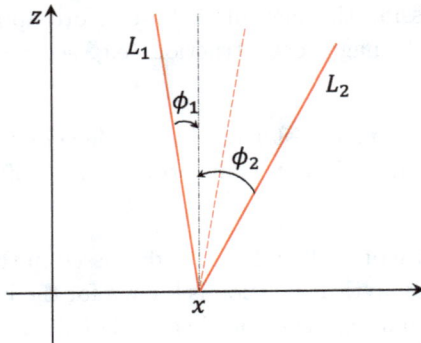

Fig. 2.27   Parametrization of V-lines by Basko *et al.* (1997a): The vertex x is restricted to the line representing the scatterer; the angles ϕ_1, ϕ_2, correspondingly between the rays L_1, L_2 and the normal to the scatter, are arbitrary.

$1/l = \cos\phi/z$, the data measured by the Compton camera at point x of the scatterer can be expressed as follows (see Figure 2.27 for a description of notations):

$$g(x, \phi_1, \phi_2) = \cos\phi_1 \int_{L_1} f(x, z)\, dl + \cos\phi_2 \int_{L_2} f(x, z)\, dl, \qquad (2.26)$$

where $z\, f(x, z)$ is the radiation source distribution supported in the upper half-space (i.e. above the scatterer).[11]

It is shown that for any fixed $K \doteq \tan\phi_1 + \tan\phi_2$, the corresponding 2D restriction of the proposed VLT of f can be used to generate the classical Radon transform of another function $\tilde{f}$, obtained from f by extending its support to the lower half-space through a mirrored shear transformation:

$$\tilde{f}(x, z) = \begin{cases} f(x, z), & z \geq 0, \\ f(x - (\tan\phi_1 + \tan\phi_2)z, -z), & z < 0, \end{cases} \qquad (2.27)$$

(see Figure 2.28 for examples of the graphical representation of this transform). Using any standard inversion technique of the classical Radon transform, one can recover $\tilde{f}(x, z)$ and thus the source distribution function $z\, f(x, z)$.

---

[11]Equation (2.26) also represents the measurements in a 2D CCI model, which ignores beam spreading but accounts for the efficiency of the detector, depending on the incidence angle ϕ of the particle. The measure of efficiency is reflected by the weight factor $\cos\phi$, and the radiation source distribution is represented by function $f(x, z)$.

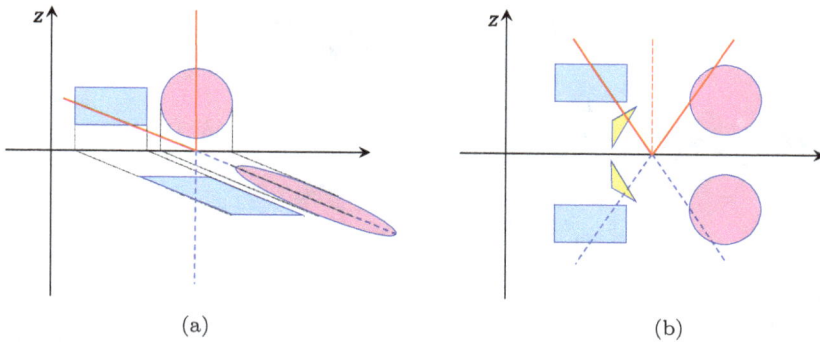

(a)   (b)

Fig. 2.28   Two graphic examples of the mirrored shear transformation described in formula (2.27). That transformation is used to generate a new function $\tilde{f}$ by extending the original function f (supported in the upper half-space) to the lower half-space. The classical Radon transform of the extended function $\tilde{f}$ can be obtained from the weighted VLT (2.26) of the original function f.

A particular case where $K = 0$ (i.e. $\hat{s}$ is normal to the sensor arrays) deserves special attention. Here, $\tilde{f}$ is obtained from f simply by an even extension with respect to variable z (see Figure 2.28(b)). As a result, many of the well-known properties of the classical Radon transform can be easily adjusted for this "vertical" VLT. In addition to 2D CCI, this transform has been also considered in relation to Compton scattering *emission* tomography with collimated receivers (Morvidone *et al.*, 2010; Truong and Nguyen, 2011a), γ-ray transmission/emission imaging (Rigaud *et al.*, 2013), and Norton's CST modality (Truong and Nguyen, 2015).

The weighted VLT, corresponding to a constant attenuation coefficient μ_t and "vertical" V-lines, was considered by Tan and Gouia-Zarrad (2017). The authors studied various properties of that attenuated VLT and established a relation between that transform and exponential Radon transform.

Remark 2.12. Almost all mathematical tasks discussed in this book are related to formally determined inverse problems associated with transforms mapping a function of d variables to a d-dimensional data set. However, in certain cases, one may be interested in overdetermined problems, where the range of the transform of interest has more degrees of freedom than that of its domain. These types of scenarios are especially common in emission tomography, where the measured data sets are severely undersampled and have a very low signal-to-noise ratio (SNR) (Wernick and Aarsvold, 2004; Allmaras *et al.*, 2013). Therefore, in such image reconstruction problems, it is crucial to use as much information as possible.

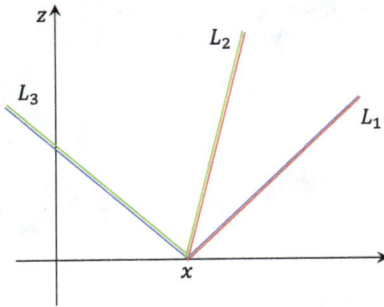

Fig. 2.29   Let f be a function supported in the upper half plane. The integral of f along the ray L_1 can be obtained by adding its integrals along V-lines L_1L_2 and L_1, L_3, subtracting its integral along V-line L_2L_3, and dividing the final result by two. Note that for each ray L_1, there exist infinitely many choices of L_2 and L_3 to perform this algorithm, so one could average over many such choices to reduce the effects of low SNR.

While the measurements in the 2D CCI modality considered by Basko *et al.* (1997a) have three degrees of freedom, the presented image reconstruction method therein uses 2D subsets of the overall data. One approach to utilize the complete data set and reduce the effects of low SNR can be averaging the reconstructions performed over many different choices of such 2D subsets.

The possibility of recovering the image in the same setup but using the entire set of measurements at once was studied by Allmaras *et al.* (2013), Hristova (2015), and Olson *et al.* (2016). The authors developed several ideas to accomplish the task and compared their efficiencies through numerical experiments. All techniques essentially boil down to generating classical Radon data of the image function by utilizing as much as possible of the VLT data. For example, one of those ideas using a finite, linear combination of VLT projections is presented in Figure 2.29. Another approach averages the data corresponding to all V-lines with a common ray to recover the divergent beam transform along that ray. The latter is equivalent to the classical Radon transform of the image function along a line containing that ray due to the restriction on the support of that function.

A more elaborate technique using the Laplace–Beltrami operator and Funk transform to recover the classical Radon data from a 3D set of VLT projections was introduced by Terzioglu (2015).[12] The feasibility of the algorithm was demonstrated through numerical simulations.

---

[12]The result presented in that paper is also applicable to more general transforms mapping functions to their integrals over cones in $\mathbb{R}^n$ for $\forall n$.

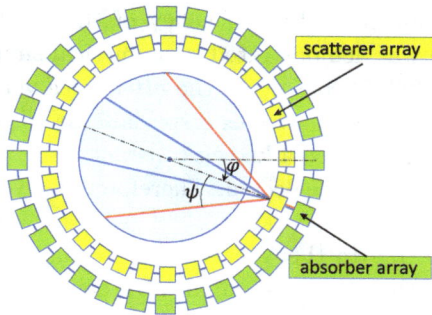

Fig. 2.30 A sketch of the 2D CCI configuration used by Moon and Haltmeier (2017), Haltmeier *et al.* (2017), and Nguyen and Nguyen (2021b). The scatterer and absorber arrays are arranged in the form of two concentric circles, with the image function located inside the disc enclosed by the scatterer. All three works consider a 2D restriction of potentially available 3D data set by considering only V-lines with an axis of symmetry passing through the center of the scanner. Therefore, the data can be parametrized by the polar angle ϕ of the V-line vertex and its half-opening angle ψ.

A different method, employing the Fourier series expansion of the data, was used by Jung and Moon (2015) to invert the full, 3D VLT with a weight function l^k for an arbitrary $k \in \mathbb{R}$. The proposed approach is similar to the one developed for the corresponding 3D Compton transform by [Parra (2000)], converting the VLT data of f into its divergent beam data and then inverting the latter through standard means (see also Section 2.5.2).

We finish this section with a discussion of 2D CCI configurations that use curvy sensor arrays. Of particular interest are setups employing a pair of concentric circular arrays as a scatterer and an absorber (see Figure 2.30). It is assumed that the image function is supported inside the smaller disc. Therefore, the measurements in this modality correspond to V-lines with vertices on the inner circle and rays directed into the smaller disc.

Moon and Haltmeier (2017) considered a 2D subset of (unweighted) VLT in this setup, restricting the V-lines of integration to only those, whose axes of symmetry pass through the center of the camera. It is easy to note that the resulting operator is rotation-invariant. Therefore, it can be diagonalized using Fourier series expansions (see Cormack, 1963; Ambartsoumian *et al.*, 2010; Ambartsoumian and Moon, 2013) for other examples of using the Fourier series to diagonalize rotation-invariant generalized Radon transforms). Moon and Haltmeier (2017) found an explicit integral relation between the nth Fourier coefficient of the image function and the nth Fourier coefficient of its VLT and used it to derive an exact analytic inversion of that transform. The efficiency of the resulting image reconstruction algorithm was demonstrated by numerical simulations.

Haltmeier *et al.* (2017) considered the *attenuated* version of the VLT in the same setup. The paper addresses a special case where the attenuation coefficient μ_t is assumed to be a known constant inside the imaging object and the beam spreading is neglected. Therefore, the weight of integration along rays of the V-lines is equal to $e^{-\mu_t l}$, where l denotes the distance from a point on the ray to the V-line vertex (recall equations (2.24) and (2.25)). It is shown that the attenuated VLT described above is injective on the space of smooth functions compactly supported inside the inner disc of the camera. Thus, the image function can be uniquely recovered from the measurements in such CCI modality. The proof uses Fourier series decompositions of the image function and its VLT data and a solution to the generalized Abel-type integral equation connecting the corresponding Fourier coefficients. The efficiency of the method is demonstrated through numerical simulations. The authors also discuss the possibility of modifying their approach to address the solution to an overdetermined inverse problem, where no restrictions are imposed on the axes of symmetry of the integration V-lines.

A recent work by Nguyen and Nguyen (2021b) studied the sampling problem for the unweighted VLT in this setup. The authors derive an explicit error estimate for recovering continuous data from its discrete samples and propose two sampling schemes fulfilling the conditions needed for the estimate. The first approach employs a standard uniform discretization of angles ϕ and ψ, representing correspondingly the polar angle of the V-line vertex position and its half-opening angle. The second one uses uniform discretization in ϕ and an interlaced discretization in ψ, resulting in a more efficient sampling scheme.

Zhang (2020a) analyzed microlocal properties of a more general VLT, integrating along a two-parameter family of V-lines with vertices on a smooth curve. An additional restriction here is that each V-line in the family can be uniquely identified by the line containing one of its branches. An example of such a set of V-lines is the collection of rays "reflecting" from a smooth boundary, i.e. V-lines with a fixed direction of axis of symmetry normal to a given smooth curve and vertices on that curve. The paper describes the notion of conjugate points of families of V-lines and shows that in the presence of conjugate points, the singularities co-normal to the broken rays cannot be recovered from local data. In other words, the image reconstructions will have artifacts if one uses VLT data limited to a small neighborhood of a particular V-line. At the same time, more singularities may be recovered from global data since in certain setups, one can probe the same singularity with different V-lines. In the special case of the VLT integrating along rays reflecting from a circle, the paper builds interesting connections with the results of Moon and Haltmeier (2017).

2.5.2 *3D models*

2.5.2.1 *Planar sensors*

Perhaps the earliest published work with an exact, analytical solution to an integral geometric inverse problem related to CCI is by Cree and Bones (1994). Their paper considers the standard 3D Compton camera setup, with a pair of planar sensor arrays serving as a scatterer and an absorber (see Figure 2.26(b)). Neglecting beam spreading and attenuation, the mathematical model presented in the article assumes that the measurements of the camera provide a five-dimensional Compton transform of the radiation source distribution. Namely, the vertices of the cones are restricted to the scatterer plane, while the directions of the axes of symmetry and the cone opening (or photon scattering) angles are arbitrary. Since the image function is 3D, it is clear that the described transform is overdetermined, and one may be able to invert its restriction to a judiciously chosen 3D subset. The authors consider such a restriction, corresponding to cones with axes of symmetry normal to the scatterer (see Figure 2.31). It is easy to see that this 3D Compton transform is invariant with respect to shifts in variables (x, y). Therefore, taking the Fourier transform of both the image function and the data with respect to those variables may simplify the task of inverting the Compton transform.

Assume that the image function $f(x, y, z)$ is smooth and compactly supported in the upper half-space $z > 0$. Let $g(x, y, t)$ be its Compton transform, corresponding to a cone with its vertex at $(x, y, 0)$ and opening angle θ, where $t = \tan \theta$. Finally, let $\mathcal{F}_2 : h(x, y) \to \hat{h}(\xi_x, \xi_y)$ denote the 2D Fourier transform in the xy-plane. It is shown by Cree and Bones (1994) that $\hat{g}(\cdot, \cdot, t)$ can be expressed through the Hankel transform or order zero of the function $\hat{f}(\cdot, \cdot, z)$. Since Hankel transform is involutive, applying it to $\hat{g}(\cdot, \cdot, t)$ and then taking an inverse 2D Fourier transform solves the problem of inverting the 3D Compton transform. Note that the theoretical result assumes that the conical integrals are known for all $(x, y) \in \mathbb{R}^2$, while in practice, the detector arrays have finite size, thus the available data set is

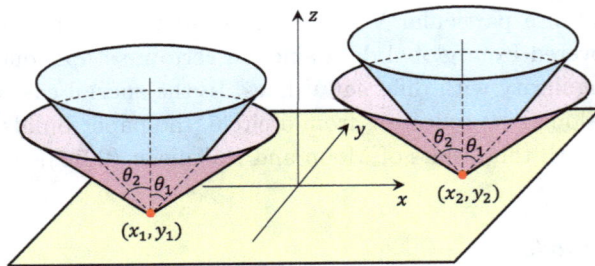

Fig. 2.31   A sketch of the 3D family of vertical cones appearing in the flat geometry of CCI. Each cone can be parametrized by its half-opening angle θ and the (x, y) coordinates of its vertex restricted to a plane.

truncated. The paper presents computer simulations testing the effects of the truncation and validating the proposed method.

Remark 2.13. The 3D Compton transform described above also naturally appears in a mathematical model of Compton scattering emission tomography (e.g. see Nguyen *et al.*, 2005). We discuss that model and some additional properties of that transform in Section 2.6.

In the discussion at the end of the article by Cree and Bones (1994), the authors outline directions for future work, arguing that "to achieve an advantage over the Anger camera, it is essential to utilize all available photon detections in the Compton camera." In other words, they too promote the objective of inverting the overdetermined transforms using as much data as possible (recall Remark 2.12). This will be a repetitive theme in the discussion of most of the works reviewed in this section.

A different approach to the problem of image reconstruction in the CCI configuration described above was developed by Basko *et al.* (1998). The authors use integrals of the image function f over cones with vertices at a point (x, y), a fixed opening angle, and arbitrary directions of the axes of symmetry (see Figure 2.32) to obtain the classical Radon data of f in $\mathbb{R}^3$ (see equation (2.31)), corresponding to planes passing through the point (x, y). After repeating this step at sufficiently many points (x, y), the image function f is reconstructed by applying a standard filtered backprojection algorithm of the Radon transform (e.g. see Natterer, 2001b). Note that this method utilizes a 4D subset of potentially available data (only the scattering angle stays fixed). The underlying mathematical machinery of the proposed technique is based on the spherical harmonics expansion of

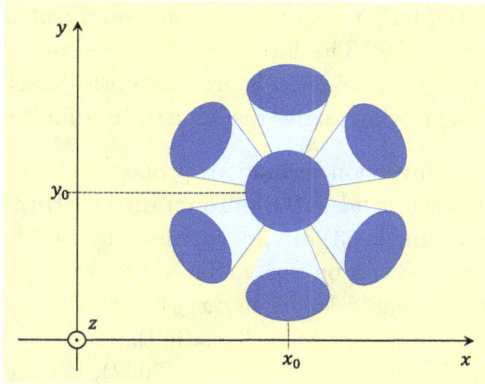

Fig. 2.32 A sketch of the 4D family of cones with fixed opening angle appearing in the flat geometry of CCI. Each cone can be parametrized by the (x,y) coordinates of its vertex restricted to a plane and the direction of its axis of symmetry.

the data. The article includes numerical simulations testing the efficiency of the developed method.

A somewhat similar methodology was presented by Parra (2000),[13] where the author employs spherical harmonics to recover the integrals of f along rays emanating from (x,y) using the integrals of f over *all* cones with a vertex at (x,y). The image function can then be recovered through standard techniques of inverting the cone beam transform (e.g. see Natterer and Wübbeling, 2001). Note that here the entire five-dimensional set of potentially available measurements is utilized. Further advancement and some generalizations of this method were later proposed in a series of works (Hirasawa *et al.*, 2001; Tomitani and Hirasawa, 2002; Hirasawa and Tomitani, 2003). In particular, it was shown that the spherical harmonics based on associated Legendre functions can be substituted with an arbitrary complete system of normalized orthogonal functions on the unit sphere. Furthermore, the authors developed a modified image

---

[13]The mathematical model of the Compton camera in this work is slightly different from the one developed by Cree and Bones (1994) and Basko *et al.* (1998), which used surface integrals of f over circular cones. Parra (2000) assumed that the count of photons corresponding to the same scattering angle and measured by a given pair of sensors on the parallel detector planes is proportional to the average of the integrals of f along rays generating the "measurement cone." The discrepancy between these two models is in the presence or absence of the distance to the vertex of the cone used as a multiplicative weight of integration.

reconstruction method that required the measurement of only a limited range of scattering angles. The latter result is desirable since, depending on the application, the photons registered by both detectors plates of the camera may have scattering angles distributed in a limited range.

Remark 2.14. The inversion techniques proposed by Basko *et al.* (1998), Parra (2000), Hirasawa *et al.* (2001), Tomitani and Hirasawa (2002), and Hirasawa and Tomitani (2003) are not intrinsically tied to the planar configuration of sensors and can be easily adapted to other geometries. Indeed, as long as the set of sensor locations (x, y) is chosen so that the planes (in the case of Basko *et al.* (1998)) or lines (in the case of Parra (2000); Hirasawa *et al.* (2001); Tomitani and Hirasawa (2002); Hirasawa and Tomitani (2003)) cover the support of the image function,[14] the proposed methodologies are applicable with little or no change.

An analytic reconstruction method of an augmented CCI model using conical surface integrals was presented by Maxim *et al.* (2009). The authors consider a Compton camera with parallel sensor plates, accounting for the sensitivity of the first detector (the scatterer) to the incidence angle of the photons.[15] The article extends the approach of Cree and Bones (1994) and shows that the radiation source function can be recovered from a 4D subset of the weighted Compton transform corresponding to cones, whose vertices are restricted to the plane of the first detector and their axes of symmetry have a fixed angle α with the normal to that plane. As with other CCI techniques that have an extra dimension of data to spare, the reconstructions over all possible choices of α can be averaged to improve the SNR ratio.

In follow-up publications (Maxim, 2014, 2019), the method presented above was given a new insight, leading to several filtered backprojection-type reconstructions of the image. The papers show a relation between the 2D Radon transforms $\mathcal{R}$ of the image function $f(x, y, z)$ in layers corresponding to fixed values of z and $\mathcal{R}$ of the weighted Compton transform as a function of the location of cone vertices on the detector plane. This leads to a closed-form, exact inversion formula for two types of 4D restrictions of the Compton transform, corresponding to a fixed opening angle of the cones or a fixed polar angle of their axes of symmetry. Using the correlation

---

[14]Such conditions in CRT inversion algorithms were later called by other authors as an *admissibility condition* (Smith, 2005; Kuchment and Terzioglu, 2016) or *Kirillov–Tuy condition* (Kuchment and Terzioglu, 2017; Terzioglu *et al.*, 2018).

[15]Recall the discussion of the 2D CCI measurements model by Basko *et al.* (1997a).

between various Compton projections, the author proposed new averaging algorithms facilitating an improvement of the statistical properties of the reconstruction.

Haltmeier (2014) introduced a filtered backprojection-type inversion formula for another weighted Compton transform, integrating the image function over surfaces of cones with vertices on a plane and axes of symmetry normal to that plane. The integration weight in this transform is an arbitrary power of the distance from the variable location on the cone to the vertex of the cone. In particular, such a weight can be used to account for the beam spreading effect. The paper also established Fourier slice identities for the aforementioned Compton transform. The results in the article are also applicable to the corresponding transforms in dimensions $d \geq 3$.

Remark 2.15. It is worth mentioning here that deriving different techniques to invert the same transform can be valuable since each approach may have its pros and cons. Moreover, while all inversion formulas must be equivalent when applied to functions located inside the range of the transform, they may produce vastly different outcomes when applied to data outside of that range. Due to noise, hardware imperfections, limited view, and other complications, the measurements done in practice typically do not satisfy the range conditions. As a result, the quest for new inversion formulas stays active even after the discovery of several methods that achieve the goal.

In the spirit of the above remark, several new studies dedicated to Compton transforms appearing in CCI with planar sensors have been published in the past couple of years. Jung and Moon (2015) considered the weighted Compton transform introduced by Haltmeier (2014) and used spherical harmonics expansions to relate that transform to the weighted divergent beam transform, akin to the approach utilized by Parra (2000), Hirasawa *et al.* (2001), Tomitani and Hirasawa (2002), and Hirasawa and Tomitani (2003). The described relation, combined with an inversion of the weighted divergent beam transform presented in the paper, leads to the recovery of the image function. The proposed method uses the full five-dimensional data set measured by the Compton camera.

Some new properties of the unweighted Compton transform, integrating over cones with axes of symmetry normal to the detector plane, are also discussed by Moon (2016) and Moon (2019).

2.5.2.2   *Flexible geometry of sensors*

In an influential article that followed soon after the works by Hirasawa and Tomitani, Smith (2005) studied both CCI models and derived exact, closed-form inversion formulas for the associated integral transforms. In the case of the model using conical surface integrals (as reported by Cree and Bones (1994) and Basko *et al.* (1998)), Smith (2005) showed that a 4D subset of the data can be used to generate the Hilbert transform $\mathcal{H}$ of the Radon transform $\mathcal{R}$ of the image function f in $\mathbb{R}^3$, leading to the subsequent recovery of f. To present that result in an explicit form, let us write down the formal definitions of the Hilbert transform and the Radon transform in $\mathbb{R}^3$.

Consider the function

$$h(t) \doteq \frac{1}{\pi t}, \tag{2.28}$$

and (with a standard abuse of notation) let h also denote the corresponding tempered distribution integrating in the sense of the Cauchy principal value (e.g. see Gelfand and Shilov, 1964):

$$\langle h, u \rangle \doteq \text{p. v.} \int_{-\infty}^{+\infty} \frac{u(t)}{\pi t} \, dt = \lim_{\varepsilon \to 0} \int_{|t| > \varepsilon} \frac{u(t)}{\pi t} \, dt, \quad \forall u \in \mathscr{S}(\mathbb{R}). \tag{2.29}$$

Then, the *Hilbert transform* of a function $g(t)$ is defined as a convolution of g with h, i.e.

$$\mathcal{H}g(t) \doteq (h * g)(t) = \frac{1}{\pi} \lim_{\varepsilon \to 0} \int_{|\tau| > \varepsilon} \frac{g(\tau)}{t - \tau} \, d\tau. \tag{2.30}$$

The *Radon transform* $\mathcal{R}f(\hat{\mathbf{n}}, t)$ of a function f in $\mathbb{R}^3$ is defined as the surface integral of f over the plane normal to the unit vector $\hat{\mathbf{n}} \in \mathbb{S}^2$ at a signed distance $t \in \mathbb{R}$ from the origin, i.e.

$$\mathcal{R}f(\hat{\mathbf{n}}, t) \doteq \int_{\mathbf{x} \cdot \hat{\mathbf{n}} = t} f(\mathbf{x}) \, d\mathbf{x} = \int_{\hat{\mathbf{n}}^\perp} f(t\hat{\mathbf{n}} + \mathbf{y}) \, d\mathbf{y}. \tag{2.31}$$

Using the above notations, the result of Smith (2005) can now be formulated as follows. Let f be a smooth function supported in the unit ball. Then,

$$\mathcal{H}\mathcal{R}f(\hat{\mathbf{s}}, \mathbf{r} \cdot \hat{\mathbf{s}}) = -\int_0^\pi \mathcal{C}f(\mathbf{r}, \hat{\mathbf{s}}, \theta) \, h(\cos\theta) \, d\theta, \tag{2.32}$$

where $\mathcal{C}f\,(\mathbf{r},\hat{\mathbf{s}},\theta)$ denotes the surface integral of f over a cone with a vertex at $\mathbf{r}$, axis of symmetry along the unit vector $\hat{\mathbf{s}}$ and half-opening angle θ, and $h(t)$ is the Cauchy kernel defined in formula (2.28). The integral on the right-hand side of (2.32) is taken in the sense of the Cauchy principal value.

The knowledge of $\mathcal{HR}f\,(\hat{\mathbf{s}},l)$, for all $\hat{\mathbf{s}}\in\mathbb{S}^2$ and $|l|\leq 1$, allows the unique recovery of f using an explicit, closed-form inversion formula. For example, let (ϕ,ψ) be the spherical angles of the unit vector $\hat{\mathbf{s}}(\phi,\psi)=(\cos\phi\,\sin\psi,\ \sin\phi\,\sin\psi,\ \cos\psi)$, and denote

$$F(\hat{\mathbf{s}},l)\doteq\frac{\partial}{\partial l}\mathcal{HR}f\,(\hat{\mathbf{s}},l). \tag{2.33}$$

Then, one can generate the 2D X-ray transforms of parallel cross sections of f by averaging $F(\hat{\mathbf{s}},l)$ over all $\hat{\mathbf{s}}$ with a fixed azimuthal angle as follows. Let $\hat{\mathbf{n}}=(\sin\phi,-\cos\phi,0)$. Then,

$$\mathcal{P}f\,(\mathrm{proj}_{\hat{\mathbf{n}}^{\perp}}\mathbf{x},\hat{\mathbf{n}})=\frac{1}{2\pi}\int_0^{\pi}F(\hat{\mathbf{s}}(\phi,\psi),\mathbf{x}\cdot\hat{\mathbf{s}}(\phi,\psi))\,d\psi, \tag{2.34}$$

where $\mathcal{P}$ denotes the X-ray transform defined in formula (1.29). Consequently, one can recover f by applying any standard inversion technique for the 2D X-ray transform (e.g. see Natterer, 2001b; Natterer and Wübbeling, 2001).

Remark 2.16. The set of formulas (2.32), (2.33), and (2.34) indicates that the reconstruction of f by this method requires at least a 4D subset of conical surface integrals of f. However, there is a built-in flexibility on the choice of an appropriate restriction of the full conical transform. In particular, one can use a subset of cones whose vertices are restricted to a curve in $\mathbb{R}^3$ but have arbitrary directions of the axis of symmetry and an arbitrary opening angle. Smith (2005) provides a rigorous description of all admissible restrictions of the full conical transform to a 4D subset through the so-called *completeness condition*.

Smith (2005) also derived a similar inversion formula for the second model of CCI, which was proposed by Parra (2000) and used later by Hirasawa *et al.* (2001), Tomitani and Hirasawa (2002), etc. Here, the count of photons $\mathcal{C}_2f\,(\mathbf{r}_s,\hat{\mathbf{s}},\theta)$, corresponding to the same scattering angle θ and measured by a given pair of sensors $\mathbf{r}_s$ and $\mathbf{r}_a$, is proportional to the average

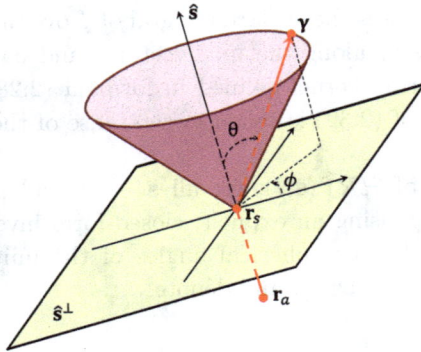

Fig. 2.33 Parametrization of rays and cones used in the second model of CCI. Here, $\mathbf{r}_s$ and $\mathbf{r}_a$ represent the locations of two point sensors, serving correspondingly as a scatterer and an absorber. Note that the sensor arrays comprising the camera are not necessarily planar. Instead, the points $\mathbf{r}_s$ and $\mathbf{r}_a$ can belong to fairly general subsets of $\mathbb{R}^3$, as long as they adhere to the completeness condition defined by Smith (2005).

of the integrals of f along rays generating the "measurement cone." More specifically, consider

$$\hat{\mathbf{s}} \doteq \frac{\mathbf{r}_s - \mathbf{r}_a}{||\mathbf{r}_s - \mathbf{r}_a||},$$

and let $\boldsymbol{\gamma}(\hat{\mathbf{s}}, \phi, \theta)$ denote the unit vector with the polar angle θ measured from $\hat{\mathbf{s}}$ and the azimuthal angle ϕ in the plane $\hat{\mathbf{s}}^{\perp}$ (see Figure 2.33). Then,

$$\mathcal{C}_2 f\left(\mathbf{r}_s, \hat{\mathbf{s}}, \theta\right) = \int_0^{2\pi} \mathcal{X} f(\mathbf{r}_s, \boldsymbol{\gamma}(\hat{\mathbf{s}}, \phi, \theta))\, d\phi, \tag{2.35}$$

where $\mathcal{X}$ denotes the divergent beam transform defined in formula (1.17). It was shown by Smith (2005) that

$$F(\hat{\mathbf{s}}, \mathbf{r} \cdot \hat{\mathbf{s}}) = \lim_{\varepsilon \to 0} \int_0^{\pi} \mathcal{C}_2 f\left(\mathbf{r}, \hat{\mathbf{s}}, \theta\right) H_{\varepsilon}(\cos \theta) \sin \theta\, d\theta, \tag{2.36}$$

where

$$H_{\varepsilon}(t) = \begin{cases} \frac{1}{\varepsilon^2}, & |t| < \varepsilon, \\ -\frac{1}{t^2}, & |t| \geq \varepsilon. \end{cases} \tag{2.37}$$

Applying formula (2.34), followed by a standard inversion of the 2D X-ray transform, completes the process of inverting $\mathcal{C}_2$.

The ideas expressed in Remark 2.16 are applicable here too. In particular, the inversion approach for $\mathcal{C}_2$ presented above can be used for a wide range of data acquisition geometries, as long as they adhere to the corresponding completeness condition outlined by Smith (2005).

An interesting generalized CRT with vertices restricted to a curve, a fixed opening angle, and an arbitrary direction of the axis of symmetry was considered by Palamodov (2017). The transform integrates the image function with a weight $w(\mathbf{x}) = |\mathbf{x}|^{1-d}$, i.e. a power of the distance from the variable location on the cone to its vertex. Note that this weight is not integrable over the whole cone. Since the dimensionality of the available data set coincides with that of the image function, the recovery of the latter is a formally determined inverse problem. Palamodov (2017) solved that problem using the inversion formula for nongeodesic Funk transform. Similar to the results of Smith (2005), here too the curve containing the vertices of integration cones should satisfy certain admissibility conditions.

Another approach to the study of the Compton transform has been used in a series of works (Terzioglu, 2015; Kuchment and Terzioglu, 2016, 2017; Terzioglu *et al.*, 2018; Terzioglu, 2019, 2020). The authors analyze the unrestricted, $2n$-dimensional CRT in $\mathbb{R}^n$ (integrating over conical surfaces with arbitrary locations of vertices, directions of axes of symmetry, and opening angles), with the goal of adapting the results to derive useful properties for various restricted versions of that transform.

Terzioglu (2015) presented a pair of integral relations between the overdetermined CRT $\mathcal{C}f$ and the classical Radon transform $\mathcal{R}f$ in $\mathbb{R}^n$, leading to a set of inversion formulas for the former. In particular, it was shown that

$$\int_0^\pi \mathcal{C}f(\mathbf{r}, \hat{\mathbf{s}}, \theta)\, d\theta = k_1(n)\, \mathcal{R}^{\#}\mathcal{R}f(\mathbf{r}), \quad \forall \hat{\mathbf{s}} \in \mathbb{S}^{n-1}, \tag{2.38}$$

where $k_1(n)$ is a constant depending on the dimension n and $\mathcal{R}^{\#}$ is the dual (backprojection) operator of the Radon transform $\mathcal{R}$ in $\mathbb{R}^n$ (recall (2.31)):

$$\mathcal{R}^{\#}g(\mathbf{r}) = \int_{\mathbb{S}^{n-1}} g(\hat{\mathbf{n}}, \hat{\mathbf{n}} \cdot \mathbf{r})\, d\hat{\mathbf{n}}. \tag{2.39}$$

An inversion formula for $\mathcal{C}$ is then easily obtained from the well-known filtered backprojection formula for $\mathcal{R}f$ (e.g. see Natterer, 2001b):

$$f(\mathbf{r}) = k_2(n)\, \mathcal{I}^{1-n} \int_0^\pi \mathcal{C}f(\mathbf{r}, \hat{\mathbf{s}}, \theta)\, d\theta, \quad \forall \hat{\mathbf{s}} \in \mathbb{S}^{n-1}, \tag{2.40}$$

where $k_2(n)$ is a known constant and $\mathcal{I}^\alpha$ denotes the Riesz potential defined by

$$\widehat{(\mathcal{I}^\alpha f)}(\xi) = |\xi|^{-\alpha}\hat{f}(\xi), \quad \alpha < n. \tag{2.41}$$

In particular, when $n = 3$, the filtering simplifies to $\mathcal{I}^{-2} = -\triangle$, where $\triangle$ denotes the Laplacian.

If $\mathcal{C}f(\mathbf{r},\hat{\mathbf{s}},\theta)$ is known for multiple directions of the axis of symmetry $\hat{\mathbf{s}} \in \mathbb{S}^{n-1}$, one can average the data over these directions in the reconstruction formula to improve the SNR. Namely, for any function $v : \mathbb{S}^{n-1} \to \mathbb{R}$ such that $\int_{\mathbb{S}^{n-1}} v(\hat{\mathbf{s}})\, d\hat{\mathbf{s}} = 1$, it follows from formula (2.40) that

$$f(\mathbf{r}) = k_2(n) \int_{\mathbb{S}^{n-1}} v(\hat{\mathbf{s}})\, \mathcal{I}^{1-n} \int_0^\pi \mathcal{C}f(\mathbf{r},\hat{\mathbf{s}},\theta)\, d\theta\, d\hat{\mathbf{s}}. \tag{2.42}$$

The second relation between $\mathcal{C}f$ and $\mathcal{R}f$ derived by Terzioglu (2015) is

$$\int_0^\pi \mathcal{C}f(\mathbf{r},\hat{\mathbf{s}},\theta)\sin\theta\, d\theta = \frac{\pi}{|\mathbb{S}^{n-1}|}\int_{\mathbb{S}^{n-1}} \mathcal{R}f(\hat{\mathbf{n}},\hat{\mathbf{n}}\cdot\mathbf{r})\,|\hat{\mathbf{n}}\cdot\hat{\mathbf{s}}|\, d\hat{\mathbf{n}}, \tag{2.43}$$

leading to the following inversion formula for $\mathcal{C}$:

$$f(\mathbf{r}) = k_3(n) \int_{\mathbb{S}^{n-1}} \mathcal{I}^{1-n} \int_0^\pi \mathcal{C}f(\mathbf{r},\hat{\mathbf{s}},\theta)\sin\theta\, d\theta\, d\hat{\mathbf{s}}. \tag{2.44}$$

In this case, the integration with respect to $\hat{\mathbf{s}}$ is a necessary part of the inversion formula and not an "artificial" averaging to improve the SNR.

Remark 2.17. Formulas (2.32) (from Smith, 2005), as well as (2.38) and (2.43) (from Terzioglu, 2015) are similar in spirit, as they average various restrictions of CRT $\mathcal{C}f$ over all possible cone opening angles to obtain an expression in terms of the Radon transform $\mathcal{R}f$. However, the averaging is done over different subsets of cones and with distinct weight functions leading to different expressions. In particular, the resulting algorithm for reconstructing f requires at least a 4D subset of CRT when using (2.32) (recall Remark 2.16), and the vertices of cones in that subset can be located outside of the support of f. In the case of formula (2.38), the resulting technique for recovering f also requires at least a 4D subset of CRT data, but the vertices of cones must be inside the support of f. The latter condition is not satisfied in CCI since the measurements therein correspond to the restriction of CRT to cones with vertices on the detector surface (outside of the support of f). Meanwhile, formula (2.43) provides more flexibility. Although the presented inversion (2.44) requires the full six-dimensional CRT data to recover f, a different approach utilizing (2.43) leads to an inversion applicable to CCI.

To describe the alternative method of reconstructing f using formula (2.43), we need the following operators.

The *cosine transform* of a function $h \in C(\mathbb{S}^{n-1})$ is defined by

$$\mathfrak{C}g(\hat{\mathbf{s}}) = \frac{1}{|\mathbb{S}^{n-1}|} \int_{\mathbb{S}^{n-1}} g(\hat{\mathbf{n}}) |\hat{\mathbf{n}} \cdot \hat{\mathbf{s}}| \, d\hat{\mathbf{n}}, \tag{2.45}$$

for all $\hat{\mathbf{s}} \in \mathbb{S}^{n-1}$.

The *translation operator* of a function $f : \mathbb{R}^n \to \mathbb{R}$ is defined by

$$\mathfrak{T}_{\mathbf{r}} f(\mathbf{x}) = f(\mathbf{x} + \mathbf{r}), \tag{2.46}$$

where $\mathbf{r} \in \mathbb{R}^n$.

Using the above notations, formula (2.43) can be rewritten as

$$\mathfrak{C}(\mathcal{R}(\mathfrak{T}_{\mathbf{r}}f))(\hat{\mathbf{s}}) = \frac{1}{|\mathbb{S}^{n-1}|} \int_{\mathbb{S}^{n-1}} \mathcal{R}(\mathfrak{T}_{\mathbf{r}}f)(\hat{\mathbf{n}}, 0) |\hat{\mathbf{n}} \cdot \hat{\mathbf{s}}| \, d\hat{\mathbf{n}}$$
$$= \frac{1}{\pi} \int_0^\pi \mathcal{C}f(\mathbf{r}, \hat{\mathbf{s}}, \theta) \sin\theta \, d\theta. \tag{2.47}$$

It is well known that the cosine transform is a continuous bijection of $C_{\text{even}}^\infty(\mathbb{S}^{n-1})$ to itself, which can be explicitly inverted using the Laplace–Beltrami operator and the Funk transform (e.g. see Gardner, 2006; Rubin, 2015). Since for any $f \in \mathscr{S}(\mathbb{R}^n)$, its Radon transform $\mathcal{R}f(\hat{\mathbf{n}}, 0) \in C_{\text{even}}^\infty(\mathbb{S}^{n-1})$, one can invert the cosine transform in formula (2.47) to recover $\mathcal{R}(\mathfrak{T}_{\mathbf{r}}f)$.

In other words, one can recover $\mathcal{R}f(\hat{\mathbf{n}}, \hat{\mathbf{n}} \cdot \mathbf{r})$ for any $\hat{\mathbf{n}} \in \mathbb{S}^{n-1}$ from the knowledge of $\{\mathcal{C}f(\mathbf{r}, \hat{\mathbf{s}}, \theta) : \forall \hat{\mathbf{s}} \in \mathbb{S}^{n-1}, 0 \le \theta \le \pi\}$ (see Terzioglu, 2015, for the details and explicit expressions). Thus, to obtain the full Radon data $\mathcal{R}f(\hat{\mathbf{n}}, t)$ for each $\hat{\mathbf{n}} \in \mathbb{S}^{n-1}$ and $t \in \mathbb{R}$ (and consequently to reconstruct f), it is sufficient to have CRT data for all cones with a common vertex at some point $\mathbf{r}$ such that $\hat{\mathbf{n}} \cdot \mathbf{r} = t$. Interestingly enough, this condition coincides with the admissibility restriction derived by Smith (2005) and discussed in our Remark 2.16.

The inversion approach described above, as well as a closely related one using a different relation between the cosine and Funk transforms, was methodically analyzed by Kuchment and Terzioglu (2016). The authors also implemented those inversion formulas numerically and tested the efficiency of the ensuing algorithms on synthetically generated data sets. The results indicated that the second approach is much more efficient than the first. A third technique based on an approximate, mollified inversion of the cosine transform was also implemented, showing promising results.

The relation between the cosine transform, CRT, and the Radon transform was also used by Terzioglu (2015) and Kuchment and Terzioglu (2016) to express the coefficients of the spherical harmonics expansion of $\mathcal{R}f$ through those of $\mathcal{C}f$. An inversion of the restricted conical transform is then accomplished by synthesizing $\mathcal{R}f$ back and inverting it through standard techniques, akin to the approach taken by Basko *et al.* (1998). This method was also numerically implemented by Kuchment and Terzioglu (2016), demonstrating good quality of image reconstruction.

Kuchment and Terzioglu (2017) presented two sets of inversion techniques for weighted CRTs in $\mathbb{R}^n$, integrating the image function with certain powers of the distance from the variable location on the cone to the vertex of the cone. The first set uses relations between weighted CRTs and weighted divergent beam transforms, for which the authors derive several novel inversion formulas. The second set is based on the relation between weighted CRTs, divergent beam transforms, and the Radon transform in $\mathbb{R}^n$. A similar inversion strategy for the unweighted CRT is analyzed by Terzioglu (2020). All the aforementioned inversion methods are applicable for a wide range of detector geometries in CCI.

Terzioglu (2019) studied some additional properties of the overdetermined CRT in $\mathbb{R}^n$ with cone vertices inside the image domain. In the case of the $n + 1$-dimensional restriction of the weighted CRT, corresponding to a fixed direction of the axis of symmetry of the cones, the article provides some generalizations of the inversion formulas 2.38 and 2.43. In the case of the full $2n$-dimensional transform, it analyzes the microlocal properties of the weighted CRT $\mathcal{C}_k$, its dual $\mathcal{C}_k^\#$, and the normal operator $\mathcal{C}_k^\# \circ \mathcal{C}_k$. The results of this paper are not directly applicable to CCI but may prove to be useful in consideration of more restricted versions of the CRT related to various imaging modalities.

A thorough study of microlocal properties of a weighted Compton transform relevant for CCI was presented by Zhang (2020b). The transform $\mathcal{C}_\kappa$

considered in the paper maps an image function f, compactly supported in $\mathbb{R}^3$, to its integrals with a smooth weight κ over conical surfaces. The vertices of the integration cones are restricted to a smooth surface away from the support of f. It is shown that when the weight function has compact support and satisfies certain nonvanishing assumptions, the normal operator $C_\kappa^\# \circ C_\kappa$ is an elliptic pseudo-differential operator at accessible singularities, which can be stably recovered from local data. The same analysis also holds for the Compton transform with vertices of cones restricted to a smooth curve and a fixed opening angle.

2.5.2.3 *Cylindrical, spherical, and conical sensor configurations*

In addition to Compton cameras comprising two planar detectors, multiple researchers have studied CCI using other geometric configurations of the sensors.

Substantial interest and efforts have been drawn to the setups where the first detector (the scatterer) is planar while the second detector (the absorber) is a nonplanar surface. Notably fruitful work on the subject was produced by investigators at the University of Michigan, who studied such cameras with the absorber in the form of a cylinder[16] (Martin *et al.*, 1993; Clinthorne *et al.*, 1996; LeBlanc *et al.*, 1998, 1999a,b; Hua *et al.*, 1999) (see Figure 2.34(a)). In a subsequent article, Chelikani *et al.* (2004) analyzed the optimal choices of the geometric parameters for Compton cameras, which utilize cylindrical or truncated conical absorbers (see Figure 2.34). The optimization parameters included the thickness, length, and diameter of the absorbing cylinder (alternatively the thickness, length, and opening angle of the absorbing truncated cone), as well as the shape, size, and location of the planar scatterer.

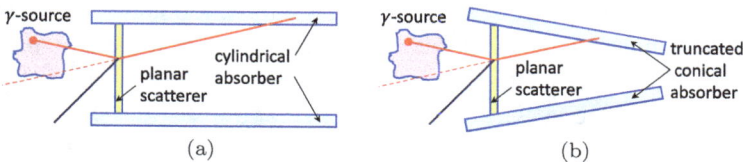

Fig. 2.34 Sketches of cross sections of ring Compton cameras made of a planar scatterer and a cylindrical (in (a)) or a truncated conical (in (b)) absorber.

---

[16]These types of devices are often called *ring Compton cameras*.

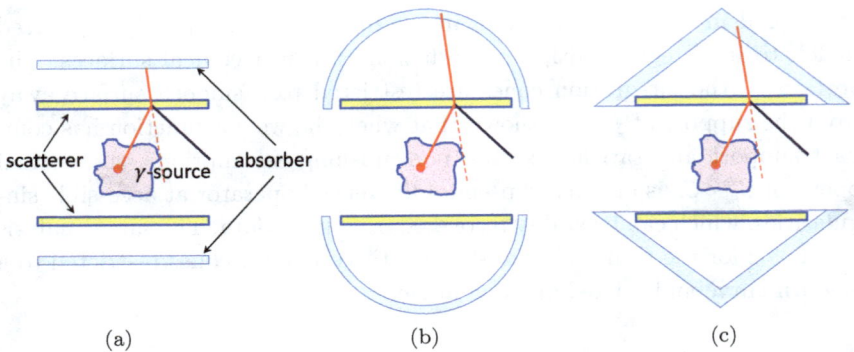

Fig. 2.35 Sketches of cross sections of double-ring Compton cameras made of a cylindrical scatterer and an absorber in the form of a surface of revolution, whose axis of symmetry coincides with that of the scatterer: (a) two concentric circular cylinders, similar to the setup proposed by Ichihara (1987), (b) used by Bolozdynya *et al.* (1997b), and (c) discussed by Rogers *et al.* (2004).

The primary emphasis of these works has been on engineering aspects of the imaging modality, including improved sensitivity, resolution, and contrast, as well as more efficient design and operation of the required hardware. Meanwhile, the mathematical models of the acquired data in all these cases still correspond to integrals along cones with vertices restricted to a plane. Therefore, the ideas related to inversion of the Compton transform described earlier are applicable here too.

Cameras employing another interesting geometry of data acquisition, where both the scatterer and the absorber are designed in the form of coaxial circular or polygonal cylinders containing the imaging object,[17] were introduced and examined in late 1980s and 1990s (Ichihara, 1987; Rohe and Valentine, 1996; Bolozdynya *et al.*, 1997b) (see Figure 2.35). Since the sensors in this setup surround the imaging object, a double-ring Compton camera has much better sensitivity than those using finite-size planar scatterers and allows complete angular sampling of the data without the need to rotate the camera. The aforementioned works did not provide exact analytic inversion formulas for the resulting Compton transform, instead recovering the image using various backprojection-type approximate inversion techniques.

Most of the analytic inversion techniques for CRT discussed in Section 2.5.2.2 can be applied to Compton transforms modeling the data acquired

---

[17]Such scanners are often called *double-ring Compton cameras.*

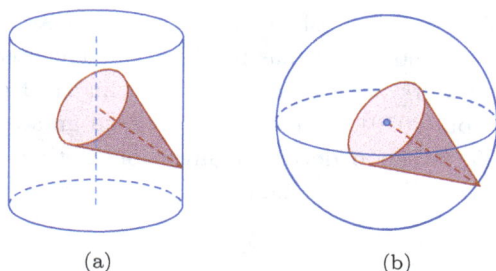

(a)   (b)

Fig. 2.36 Sketches of cones with vertices on the surfaces of a scatterer. Cones in (a) have four degrees of freedom: two parameterizing the location of the vertex on the cylinder, one describing the angle between the axes of symmetry of the cylinder and the cone, and the last one representing the cone's opening angle. Cones in (b) have three degrees of freedom: two parameterizing the location of the vertex on the sphere and one representing the cone's opening angle.

by double-ring cameras. At the same time, one may hope that more efficient approaches can be derived by utilizing the symmetries of the cylindrical geometry. Several results in this direction have been published in the past couple of years.

Moon and Haltmeier (2017) presented two analytic inversion methods for a 4D restriction of the Compton transform pertinent to the double-ring geometry. The restricted transform corresponds to integrals over cones with vertices on a circular cylinder and symmetry axes intersecting that of the cylinder (see Figure 2.36(a)). The first approach is based on reducing the Compton transform to an associated V-line transform in a disc and inverting the latter (recall our discussion of Moon and Haltmeier (2017) in Section 2.5.1). This inversion technique is also applicable to the weighted Compton transform, integrating the image function with an integer power of the distance from the variable location on the cone to the vertex of the cone. The second method follows the approach taken by Smith (2005), reducing the Compton transform to the classical Radon transform. The authors then use this technique to derive explicit formulas and Sobolev stability estimates for inversion of the 4D Compton transform.

Moon (2017) considered a more general model, in which the Compton transform is five-dimensional, integrating over cones with vertices on a surface of revolution around a fixed axis. The integrals also have a weight factor similar to the one considered by Moon and Haltmeier (2017) discussed above. The proposed technique of inverting the transform is rather involved, requiring successive inversions of the spherical sectional and the weighted fan beam transforms and, so far, has not been implemented numerically.

Another recent work associated with the cylindrical geometry of the scatterer is by Kwon (2019). Here, the author used an adaptation of ideas from Smith (2005, 2011) to derive an inversion method for a Compton transform integrating the image function along cones with vertices on a helical curve.

A different, compelling design of a scanner with inherent symmetries is the Compton camera that uses concentric spherical detectors. Although a complete spherical geometry of data acquisition is not practical, a truncated version with a spherical cap opening for the patient can be suitable in clinical applications (e.g. for brain imaging). Various properties of Compton cameras with spherical sensors have been considered in the 1990s (e.g. see Rohe and Valentine, 1996; Rohe *et al.*, 1997; Valentine *et al.*, 1997), including approximate inversions of the associated Compton transform using the point spread function of its 3D backprojection.

More recently, Schiefeneder and Haltmeier (2017) studied the inversion of a 3D restriction of the Compton transform integrating over cones with vertices restricted to a sphere and symmetry axes passing through its center (see Figure 2.36(b)). The authors derived an analytic inversion technique for that transform using spherical harmonic decompositions and solved the resulting 1D integral equations of generalized Abel type. The efficiency of the proposed method was demonstrated through numerical examples.

An ensuing work (Haltmeier and Schiefeneder, 2018) examined a weighted Compton transform in the same geometry. Here, the weight is an arbitrary smooth function of the distance from the variable location on the cone to the vertex of the cone. It is shown that the weighted transform is invertible, albeit the inversion is ill-posed. To address the latter issue, the authors apply a convex variational regularization and present numerical studies comparing various regularizers.

Another interesting and efficient design of a scanner with curved detector surfaces was proposed by Smith (2011). The Compton camera here comprises two coaxial conical surfaces serving as a scatterer and an absorber (see Figure 2.37). Image reconstruction from the measurements of this "lampshade" camera is achieved by using inversion techniques of two different Compton transforms described by Smith (2005). Numerical simulations of the proposed model were presented by Smith (2012).

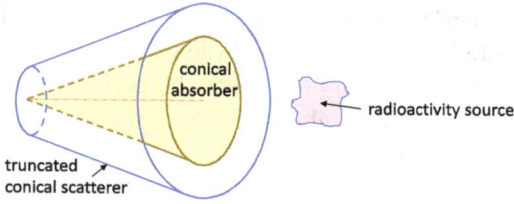

Fig. 2.37 A sketch of the "lampshade" Compton camera proposed by Smith (2011).

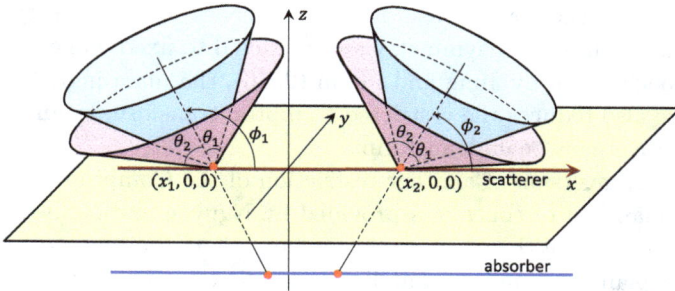

Fig. 2.38 A set of cones with three degrees of freedom, corresponding to the data measured by a Compton camera with a linear scatterer and a linear absorber.

2.5.2.4 *Linear sensor configurations*

In this section, we discuss several articles on Compton transforms with cone vertices restricted to a line. The possibility and various techniques of inverting such transforms were also discussed in Section 2.5.2.2. It is clear that the sensitivity of 1D cameras imaging a 3D source will be poor. Therefore, the results of these works have to be enhanced or integrated with additional techniques to be applicable to 3D CCI.

Jung and Moon (2016) considered a 3D restriction of the Compton transform integrating over cones with vertices on a line, axes of symmetry in a fixed plane, and an arbitrary opening angle (see Figure 2.38). It is easy to see that this transform has a nontrivial kernel, including all functions that are odd with respect to the plane containing the axes of symmetry of the cones. The authors show in the paper that the transform is injective on the space of smooth functions compactly supported on one side of the plane and derive an exact inversion formula for that case. The result is obtained by formulating a relation between the Compton transform and the weighted divergent beam transform and inverting the latter.

An impediment of the inversion formula presented by Jung and Moon (2016) is its use of Compton data along the entire line, i.e. one would need to have infinitely long detectors to collect the necessary information. This drawback was eliminated in a follow-up paper (Moon and Haltmeier, 2020), showing that one can uniquely recover the image function from the data collected along three mutually orthogonal linear detectors of finite size. Namely, they considered the restriction of the Compton transform to the following 3D submanifold of cones. The vertices of cones are located on the union of line segments $\{c\mathbf{e}_j\}$, $j = 1, 2, 3$, where $c \in [0, 1]$ and $\mathbf{e}_j$s represent the standard orthonormal basis in $\mathbb{R}^3$, the cones have arbitrary opening angles, and their axes of symmetry are restricted to fixed planes. Similar to the approach taken by Jung and Moon (2016), the main ingredient of the proof here also reduces the restricted Compton transform to an associated weighted divergent beam transform.

An exact inversion of another restriction of the Compton transform to a 3D submanifold of cones was provided by Nguyen and Nguyen (2021a). Here, the vertices of the cones are located on a line, their axes of symmetry belong to planes normal to that line, and the opening angle is arbitrary (see Figure 2.39). One possible realization of such setups, in practice, can be accomplished through the rotation of a pair of linear detectors around the axis coinciding with the scatterer. The inversion formula makes use of the vertical slice transform on a sphere and the VLT on a plane.

2.5.2.5 *Compton cameras with multiple layers and VLT in 3D*

Since the early years of the development of Compton cameras, various researchers have studied CCI models comprising multiple detector layers

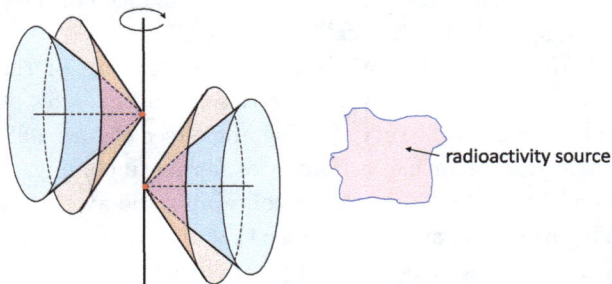

radioactivity source

Fig. 2.39   A set of "horizontal" cones with three degrees of freedom. The integrals over these cones correspond to the data measured by a Compton camera, which consists of a linear absorber and a linear scatterer and is rotated around the latter.

(a stack of scatterers and one absorber). The goals of the initial works on
the subject were twofold: improving the detection efficiency (Kamae *et al.*,
1987, 1988; Dogan *et al.*, 1990) and reducing the directional ambiguity of
incoming photons (Dogan *et al.*, 1992; Basko *et al.*, 1997b; Chelikani *et al.*,
2004).

The premise of the latter is rooted in the physical phenomenon of
polarization of photons undergoing Compton scattering (e.g. see Wightman, 1948; Fernández *et al.*, 1993; Fernández, 1999, and the references
therein). In simple terms, that phenomenon can be described as follows.
When a γ-ray scatters twice, the first event polarizes the scattered photons, inducing the preferred azimuthal direction of subsequent scattering.
As a result, the scattering and absorption positions of the double-scattered
photons and the point source of radiation tend to be located on (or close
to) the same plane (Basko *et al.*, 1997b; Chelikani *et al.*, 2004). Therefore,
the data measured by a Compton camera comprising three detector layers
can be approximated by a V-line transform of the image function in 3D
(see Figure 2.40).

The manifold of V-lines in $\mathbb{R}^3$ with vertices on a plane (say $z = 0$) has
six degrees of freedom. Naturally, it is expected that one should be able to
invert its restrictions to various submanifolds of dimension $d \geq 3$.

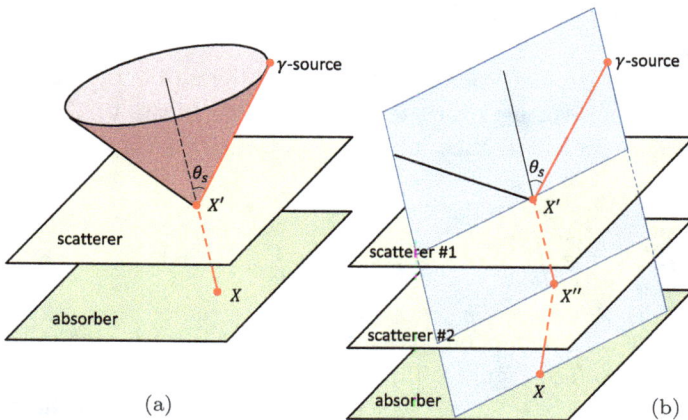

Fig. 2.40   A sketch of potential locations of the γ-source based on photon scattering
and absorption positions: In (a), one scattering and one absorption position induce a
cone of potential locations. In (b), two scattering and one absorption position generate a
V-line (intersection of a cone and a plane). An alternative design of the multiple-scatter
Compton camera uses a cylindrical absorber surrounding the stack of scatterers (not
pictured).

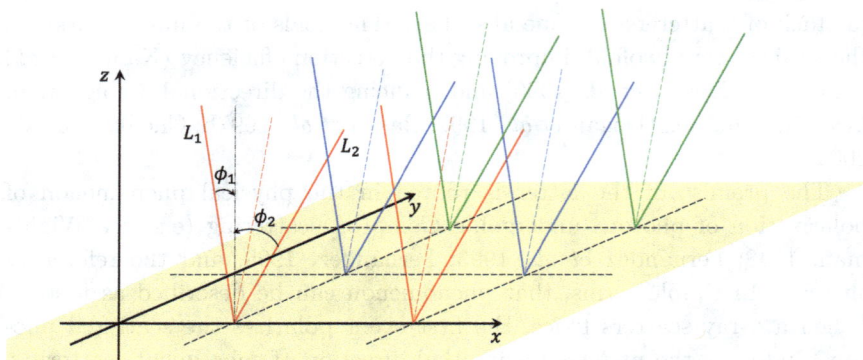

Fig. 2.41   A sketch of a family of V-lines corresponding to an invertible 3D VLT.

One obvious choice of a restriction for $d = 3$ is the following. Consider the V-lines in the xz-plane with vertices on the x-axis (see Figure 2.41). Assume also that they satisfy the constraint $\tan \phi_1 + \tan \phi_2 = \text{const}$, described before formula (2.27). Extend the family of V-lines by translation in the direction of y-axis to a 3D set of V-lines, with vertices at every point of the xy-plane. Then, employing the inversion technique from Basko *et al.* (1997a) presented in Section 2.5.1, one can use that 3D VLT data to recover the 3D image function slice by slice, i.e. separately in each plane parallel to the xz-plane.

As in other configurations of CCI (recall Remark 2.12), one may also be interested in inverting overdetermined versions of the VLT in $\mathbb{R}^3$. Basko *et al.* (1997b) presented such an inversion formula for a 4D restriction of the VLT:

$$p(\mathbf{r}, \mathbf{k}', \mathbf{k}'') = \int_0^\infty f(\mathbf{r} + \mathbf{k}'z, z)\, dz + \int_0^\infty f(\mathbf{r} + \mathbf{k}''z, z)\, dz, \qquad (2.48)$$

where $\mathbf{r}, \mathbf{k}', \mathbf{k}''$ are vectors on the xy-plane and f is supported in the upper half-space $z \geq 0$ (see Figure 2.42).

In particular, it was shown that for any fixed vector $\mathbf{K}$ in the xy-plane, the 4D VLT restriction $p(\mathbf{r}, \mathbf{k}, \mathbf{K} - \mathbf{k})$ coincides with the X-ray transform of the modified image function

$$f_{\mathbf{K}}(\mathbf{r}, z) = \begin{cases} f(\mathbf{r}, z), & z \geq 0, \\ f(\mathbf{r} - \mathbf{K}z, -z), & z < 0. \end{cases} \qquad (2.49)$$

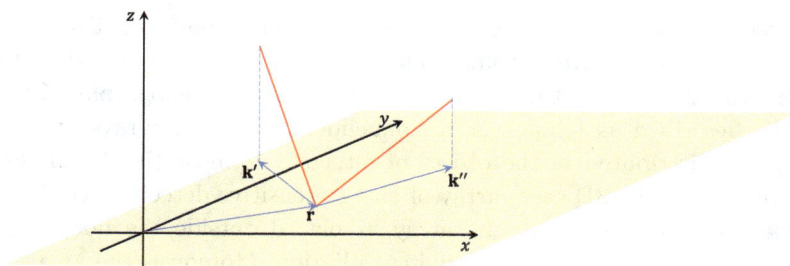

Fig. 2.42   Parametrization of V-lines in (Basko *et al.*, 1997b).

In other words,

$$p(\mathbf{r}, \mathbf{k}, \mathbf{K} - \mathbf{k}) = \int_{-\infty}^{\infty} f_{\mathbf{K}}(\mathbf{r} + \mathbf{k}z, z) \, dz. \tag{2.50}$$

Using standard techniques of inverting the X-ray transform in $\mathbb{R}^3$ (e.g. see Natterer, 2001b), one can recover $f_{\mathbf{K}}$, which coincides with f in the upper half plane.

Remark 2.18. Since the inversion algorithm described above works for an arbitrary choice of vector $\mathbf{K}$, one can utilize this redundancy to reduce the effects of low SNR in CCI by averaging the reconstructions performed over many different choices of $\mathbf{K}$.

There are many other interesting papers dedicated to multiple-scattering Compton cameras, primarily concerned with various engineering aspects of the technology. One prominent application is its use for the detection and simultaneous imaging of different γ-ray emitting radionuclides in a wide energy range. Since these works do not use integral geometric models, we do not discuss them here and refer the interested reader to the appropriate literature (e.g. see Bolozdynya and Morgunov, 1998; Kroeger *et al.*, 2002; Wulf *et al.*, 2004, 2007; Kim and Lee, 2008; Motomura *et al.*, 2008; Richard *et al.*, 2010; Seo *et al.*, 2010; Kim, 2018, and the references therein).

2.6   Compton Scattering Emission Tomography

In this section, we discuss *emission* tomography techniques that use Compton scattered photons and *mechanically collimated* receivers. To emphasize the difference from previously considered modalities, we recall that

Section 2.4 was dedicated to transmission tomography and Section 2.5 examined imaging systems using omnidirectional (uncollimated) detectors.

A typical model of Compton scattering emission tomography (CSET) can be described as follows. A radionuclide, generating γ-rays of known energy, is distributed in the object of interest. A linear (in the 2D case) or a planar (in the 3D case) array of energy-sensitive detectors, collimated in the direction normal to the array, is placed outside the body, which measures the intensity of the arriving radiation. Moreover, the arrays are oriented so that the detectors do not "directly see" the object, i.e. they do not register any ballistic photon. Finally, it is assumed that all photons registered by these detectors have scattered exactly once inside the object. This simplification is justified by the dominance of Compton single-scatter events in the interaction of γ-rays with the medium (Zaidi, 1999; Zaidi and Koral, 2004).

When a given detector in the array measures the energy of an incoming photon, it is used in conjunction with the Compton scattering relation (1.12) to determine the single-scattering angle θ of that photon. As a result, the set of feasible flight trajectories of that photon is limited to V-lines with an opening angle $\pi - \theta$, while one of its rays is passing through the detector location and is normal to the detector array (see Figure 2.43).

Due to directional ambiguity, the set of potential locations of γ-sources, generating photons that single-scatter at the same point $\mathbf{M}$ and arrive at

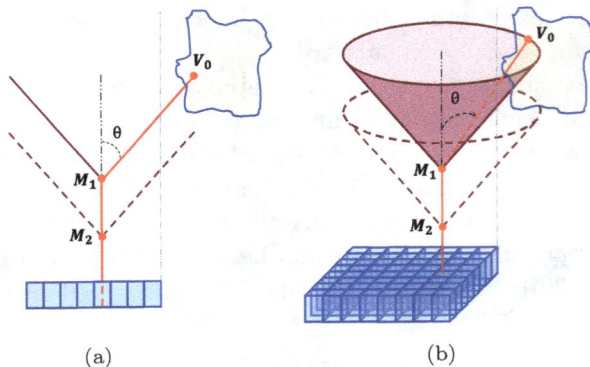

(a)   (b)

Fig. 2.43   The sets of feasible flight trajectories of a photon based on the energy of the photon measured by a detector and the position of that detector in 2D (a) and in 3D (b).

the same detector with the same energy, constitutes a vertical[18] V-line (in 2D) or a vertical cone (in 3D) with its vertex at $\mathbf{M}$.

Taking into account the beam spreading effect and noting that the scattering point $\mathbf{M} = (x_d, y_d, z_m)$ acts as a source of secondary emission, the measurement of a detector located at the point $\mathbf{D} = (x_d, y_d, 0)$ can be mathematically modeled as

$$g(\mathbf{D}, \theta) = \mathcal{T}_c f(\mathbf{D}, \theta) \tag{2.51}$$

$$\doteq K(\theta) \int_0^\infty \frac{1}{z_m{}^2} \int_0^{2\pi} \int_0^\infty f(x_d + r \sin\theta \cos\phi, y_d$$

$$+ r \sin\theta \sin\phi, z_m + r \cos\theta) \frac{1}{r} dr \, d\phi \, dz_m,$$

where $K(\theta)$ is a known function and the scattering angle θ is defined by the measured energy level. For a detailed derivation of the above relation, see Nguyen and Truong (2002) and Nguyen *et al.* (2004).

The convergence of integrals in equation (2.51) follows from the proper choice of the location and orientation of the detectors, ensuring that the image function f is supported away from the line/plane containing the detector array and the vertical space above it.

Note that the interior double integral in (2.51) corresponds to a weighted CRT of the image function, integrating over the vertical cone with a vertex at $\mathbf{M}$. Meanwhile, the exterior integration sums up (with a weight) the CRT data over all possible scattering locations $\mathbf{M}$. The transform $\mathcal{T}_c$ was first introduced by Nguyen and Truong (2002) and referred to by Nguyen *et al.* (2004) and in subsequent papers as *compounded conical Radon transform* (*CCRT*).

Nguyen and Truong (2002) studied various properties of CCRT and showed that it is invertible. A follow-up work (Nguyen *et al.*, 2004) introduced an inversion procedure for CCRT using the so-called *scatter point spread function* and demonstrated its efficiency on several numerical examples. Some additional simulations using the latter approach were presented by Nguyen *et al.* (2009).

Nguyen *et al.* (2005) took a different approach to the inversion of CCRT. Changing the order of integration in Definition (2.51), the function g can

---

[18]We call a V-line/cone *vertical* if its axis of symmetry is normal to the detector array.

be expressed as

$$g = \mathcal{T}_c f = \mathcal{C}_1 h, \tag{2.52}$$

where $\mathcal{C}_1$ is the 3D Compton transform integrating over vertical cones with vertices on the plane of the detector and h is a weighted divergent beam transform of f, namely:

$$h(x, y, z) = \int_0^\infty \frac{1}{z_m{}^2} f(x, y, z_m + z) \, dz_m. \tag{2.53}$$

Since h uniquely identifies f, the problem of inverting $\mathcal{T}_c$ boils down to the inversion of $\mathcal{C}_1$, which can be accomplished, for example, by using the approach taken by Cree and Bones (1994) (discussed in Section 2.5.2.1 in relation to 3D CCI with planar sensors).

In addition to a thorough description of an inversion procedure for $\mathcal{C}_1$, Nguyen *et al.* (2005) studied in great detail various properties of that transform, including its relations with translation, rotation, scaling, and certain differentiation operators. They established an analogue of the Plancherel formula for $\mathcal{C}_1$, derived its Schwartz kernel, as well as an alternative form of the transform expressed through a double Hankel transform for circular components. Some examples of $\mathcal{C}_1$ acting on simple functions were explicitly calculated and characterized in terms of the classical Radon transform.

In a more recent article on $\mathcal{C}_1$ in the context of CSET, Cebeiro *et al.* (2016) scrupulously discussed the features of the backprojection of data corresponding to $\mathcal{C}_1$, as well as the associated filters from its exact inversion formula established by Haltmeier (2014). The paper also presents extensive numerical simulations of the forward operator and its filtered backprojection inversion.

A novel microlocal analysis of generalized Radon transforms integrating over the surfaces of revolution of smooth curves was presented by Webber and Quinto (2021). It was shown that the results of the study are applicable to the CSET, establishing an *a priori* characterization of visible singularities and added artifacts in a reconstructed image. The paper provided simulated image reconstructions using various types of phantoms. The artifacts predicted in the theoretical findings coincided with the actual ones that appeared in the reconstructions.

The ideas about image reconstruction in 3D CSET discussed above can be essentially repeated for the 2D case, leading to consideration of the *compounded V-line transform (CVLT)* (Nguyen *et al.*, 2011). An inversion formula for CVLT then follows from the inversion of a 2D VLT integrating along vertical V-lines with vertices on a line. To accomplish the latter, various analytic and algebraic inversion techniques have been developed by Morvidone *et al.* (2010), Truong and Nguyen (2011a), and Cebeiro and Morvidone (2013), including filtered backprojection-type formulas and SVD inversions.

Another potential technique to solve this problem is the one developed by Basko *et al.* (1997a). Using an even extension $\tilde{f}$ of the original image function f across the line of the detector array, one can reformulate the problem of inverting the VLT acting on f to a problem of inverting the classical Radon transform acting on $\tilde{f}$. Consequently, the entire machinery developed for the classical Radon transform becomes available, including multiple inversion formulas and algorithms.

Remark 2.19. The CSET modality described above uses mechanical collimation, which discards the majority of scattered photons. Therefore, it is natural to contemplate the use of uncollimated detectors (either single-layer energy-resolved detectors or Compton cameras) to register photons, which have single-scattered inside the image domain. The mathematical model of data acquired by a detector in such a device will correspond to much more complicated generalized CCRTs or CVLTs (e.g. see Nguyen *et al.*, 2009, and the references therein). Analytical inversions of such transforms are still open problems. A potential step in the direction of solving some of them may be the work by Truong *et al.* (2007), where the authors derived an inversion for a 4D (overdetermined) CRT with unrestricted location of vertices, radial axis of symmetry, and an arbitrary opening angle.

2.7 Additional References

- The possibility of using *single-scattered radiation in transmission tomography* goes back to at least the 1950s. The imaging systems proposed in the early works did not use integral geometric models, and their applications were hampered by the lack of sophisticated instrumentation and powerful computational resources. Nevertheless, many of the main physical ideas have already been outlined there. An excellent account of the history of single-scattered transmission tomography and a survey of relevant works can be found in the review article by Truong and Nguyen (2012). In the following, we list the significant publications that predated the introduction of models using generalized Radon transforms and were not cited in this chapter. While we made an effort to acknowledge as many of the influential contributions to the subject as possible, the list is certainly not comprehensive, and we may have accidentally missed some important papers:

Lale (1959, 1968); Clarke (1965); Clarke and Van Dyk (1969, 1973); Kondic and Hahn (1970); Farmer and Collins (1971, 1974); Reiss and Schuster (1972); Garnett *et al.* (1973); Dohring *et al.* (1974); Clarke *et al.* (1976); Webber and Kennett (1976); Battista *et al.* (1977); Kondic (1978); Battista and Bronskill (1978, 1981); Huddleston and Bhaduri (1979); Towe and Jacobs (1981); Harding (1982); Harding *et al.* (1983); Gautam *et al.* (1983); Kondic *et al.* (1983); Holt *et al.* (1983, 1984); Brateman *et al.* (1984); Bodette and Jacobs (1984); Holt (1985); Harding and Tischler (1986); Hussein *et al.* (1986); Jacobs (1986); Guzzardi and Licitra (1987); Holt and Cooper (1987, 1988); Berodias and Peix (1988); Harding and Kosanetzky (1989); Anghaie *et al.* (1990); Speller and Horrocks (1991); Balogun and Spyrou (1993); Prettyman *et al.* (1993); Arendtsz and Hussein (1993, 1995a,b); Harding (1997).

- In addition to its use in numerous areas of medicine and biology (see the references listed earlier as well as more recent works (Lewis *et al.*, 2000; Sharaf, 2001; Cong and Wang, 2011; Alpuche Aviles *et al.*, 2011; Lee and Chen, 2015; Jones *et al.*, 2018; Redler *et al.*, 2018)), single-scattering tomography also has a broad range of other applications, such as measurement of fluid density in pipes (Kondic and Hahn, 1970; Kondic, 1978; Kondic *et al.*, 1983), soil imaging (Balogun and Cruvinel, 2003; Cruvinel and Balogun, 2006; Adejumo *et al.*, 2011), industrial nondestructive testing (Holt and Cooper, 1988; Anghaie *et al.*, 1990; Balogun and Spyrou, 1993; Evans *et al.*, 1998, 1999, 2002; Cesareo *et al.*, 2002; Brunetti *et al.*, 2002; Achmad and Hussein, 2004; Gorshkov *et al.*, 2005; Arsenault and Hussein, 2006; Hussein, 2007; Grubsky *et al.*, 2011; Kolkoori *et al.*, 2015), imaging in archaeology (Harding and Harding, 2010; Prado *et al.*, 2017), and homeland security (Hussein and Waller, 1998; Hussein *et al.*, 2005).
- In the following, we list the significant publications on *CCI* that did not concern integral geometry and were not cited in this chapter. Most of them are dedicated to the engineering and/or clinical aspects of the modality. While we made an effort to acknowledge as many of the influential contributions to the subject as possible, the list is certainly not comprehensive, and we may have accidentally missed some important papers:

Solomon and Ott (1988); Singh *et al.* (1988); Singh and Brechner (1990); Schönfelder *et al.* (1993); King *et al.* (1994); Martin *et al.* (1994); McKisson *et al.* (1994); Singh *et al.* (1995); Bolozdynya *et al.* (1997a); Gormley *et al.* (1996); Royle and Speller (1996, 1997); Ordonez *et al.* (1997a); Tumer *et al.* (1997); Wilderman *et al.* (1997); Valentine *et al.* (1997); Leblanc (1999); Ordonez *et al.* (1999); Antich *et al.* (2000); Boggs and Jean (2000); Earnhart *et al.* (2000); Hua (2000); Scannavini *et al.* (2000); Li *et al.* (2001); Yang *et al.* (2001); Conka-Nurdan *et al.* (2002); Zoglauer and Kanbach (2003); Kurfess *et al.* (2004); Zhang *et al.* (2004); Boggs *et al.* (2005); Watanabe *et al.* (2005); Lackie *et al.* (2006); Xu and He (2006); Hill and Matthews (2007); Kim *et al.* (2007); Han *et al.* (2008); Hruska *et al.* (2008); Sinclair *et al.* (2009); Frandes *et al.* (2010); Lee and Lee (2010); Lingenfelter *et al.* (2010); Kurosawa *et al.* (2012); Uche *et al.* (2012); Maxim *et al.* (2015); Smith (2015); Takeda *et al.* (2015); Wahl *et al.* (2015); Hilaire *et al.* (2016); Aldawood *et al.* (2017); Fontana *et al.* (2017); Kishimoto *et al.* (2017); Muñoz *et al.* (2018); Tashima *et al.* (2020).

- Various groups have recently worked on the development of neutron detectors based on the neutron double-scatter technique, generating conical transform data similar to that in CCI (Hutcheson and Phlips, 2009; Spence, 2011; Goldsmith *et al.*, 2016; Folsom, 2020).

PART II
Generalized Radon Transforms
"With a Vertex"

Chapter 3

V-line and Conical Radon Transforms in Slab Geometry

3.1 Definitions

We start this chapter with formal definitions of the relevant generalized Radon transforms. Although they have appeared in one form or another during the discussion of the RTE and imaging modalities in prior chapters, we list them here again for the convenience of the reader.

In all the following definitions, it is assumed that the function f decays sufficiently fast at infinity so that the corresponding integrals converge. For simplicity, one can assume that $f \in \mathscr{S}(\mathbb{R}^d)$, the Schwartz space, or $f \in C_c^\infty(\mathbb{R}^d)$, the space of compactly supported smooth functions.

Definition 3.1. The *divergent beam transform* maps a function $f : \mathbb{R}^d \to \mathbb{R}$ to its integrals along rays emanating from a point $\mathbf{r} \in \mathbb{R}^d$ in the direction $\hat{\mathbf{n}} \in \mathbb{S}^{d-1}$. More formally (recall formula (1.17)),

$$\mathcal{X}f(\mathbf{r}, \hat{\mathbf{n}}) \doteq \int_0^\infty f(\mathbf{r} + l\hat{\mathbf{n}}) \, dl.$$

In the general case, $\hat{\mathbf{n}}$ can be a vector field, i.e. $\hat{\mathbf{n}} = \hat{\mathbf{n}}(\mathbf{r})$.

Definition 3.2. The *V-line transform* (*VLT*) (sometimes also called *broken ray transform* (*BRT*)) maps a function $f : \mathbb{R}^d \to \mathbb{R}$ to a linear combination of its divergent beam transforms emanating from a common point $\mathbf{r} \in \mathbb{R}^d$ in the directions $\boldsymbol{\gamma}_1$ and $\boldsymbol{\gamma}_2$. More formally (recall formula (1.42)),

$$\mathcal{B}^{\mathbf{c}}f(\mathbf{r}, \Gamma) \doteq c_1(\mathcal{X}f)(\mathbf{r}, \boldsymbol{\gamma}_1) + c_2(\mathcal{X}f)(\mathbf{r}, \boldsymbol{\gamma}_2), \quad \mathbf{r} \in \mathbb{R}^d,$$

where $\mathbf{c} = (c_1, c_2) \in \mathbb{R}^2$ is such that $c_1 c_2 \neq 0$ and $\Gamma = [\boldsymbol{\gamma}_1, \boldsymbol{\gamma}_2]$ is a $d \times 2$ matrix with columns $\boldsymbol{\gamma}_j \in \mathbb{S}^{d-1}$, $j = 1, 2$, which also may depend on $\mathbf{r}$.

Definition 3.3. The *conical Radon transform* (*CRT*) maps a function f : $\mathbb{R}^d \to \mathbb{R}$ to its surface integrals over cones $C(\mathbf{r}, \hat{\mathbf{n}}, \beta)$ with a vertex at $\mathbf{r} \in \mathbb{R}^d$, the axis of symmetry in the direction $\hat{\mathbf{n}} \in \mathbb{S}^{d-1}$, and half-opening angle $\beta \in (0, \pi/2)$. More formally (recall formula (2.25)),

$$\mathcal{C}f(\mathbf{r}, \hat{\mathbf{n}}, \beta) \doteq \int\limits_{C(\mathbf{r}, \hat{\mathbf{n}}, \beta)} f(\mathbf{r}) \, ds.$$

SSOT in slab geometry requires a formally determined inversion of VLT with uniform weights, arbitrary locations of vertices inside the image domain, a fixed opening angle, and fixed ray directions (see Remark 2.1). For simplicity of computations, we choose the following parametrization: $\mathbf{r} = (x_v, y_v)$, $c_1 = c_2 = 1$, $\boldsymbol{\gamma}_1 = (-\cos\beta, \sin\beta)$, and $\boldsymbol{\gamma}_2 = (\cos\beta, \sin\beta)^1$ (see Figure 3.1(a)). Under the above assumptions, we denote

$$\mathcal{B}_\beta f(\mathbf{r}) \doteq \mathcal{B}^c f(\mathbf{r}, \Gamma). \tag{3.1}$$

Similarly, in the case of CRT we consider a three-dimensional restriction corresponding to arbitrary locations of vertices $\mathbf{r} = (x_v, y_v, z_v)$ inside the image domain, a fixed half-opening angle β, and a fixed axis of symmetry along the vector $\hat{\mathbf{n}} = (0, 0, 1)$ (see Figure 3.1 (b)). Under the above assumptions, we denote

$$\mathcal{C}_\beta f(\mathbf{r}) \doteq \mathcal{C}f(\mathbf{r}, \hat{\mathbf{n}}, \beta). \tag{3.2}$$

3.2   Fourier Inversion

It is easy to note that the VLT and CRT described above are translation-invariant. A common method for inverting such operators is to apply the Fourier transform, which diagonalizes the operator, and then solve the resulting simpler equation in the Fourier domain (e.g. see Kuchment, 2013, Chapter 15).

3.2.1   *V-line transform*

The first exact analytic inversion of the VLT was obtained using Fourier techniques by Florescu *et al.* (2011). A simpler inversion formula using

---

[1]To avoid cumbersome notation in formulas, we will not use the transpose symbol ^{T} to distinguish between the row and column vectors.

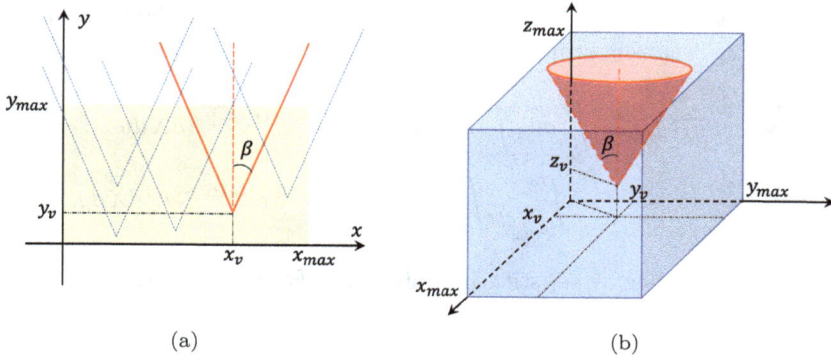

Fig. 3.1   Parametrization of V-lines (a) and (circular) cones (b) in slab geometry.

Fourier techniques was derived later by Gouia-Zarrad and Ambartsoumian (2014). In the following, we present the result from the latter work.

Theorem 3.1. *Consider a function $f \in C_c^\infty(\mathbb{R}^2)$ supported in a rectangle $R = \{(x, y) \in \mathbb{R}^2 \mid 0 \le x \le x_{max}, 0 \le y \le y_{max}\}$. For each vertex location $(x_v, y_v) \in \mathbb{R}^2$, denote the VLT of f by*

$$g(x_v, y_v) = \mathcal{B}_\beta f(x_v, y_v). \tag{3.3}$$

An exact inversion of the VLT is given by the formula

$$f(x, y) = -\frac{\cos \beta}{2} \left[\frac{\partial}{\partial y} g(x, y) + \tan^2 \beta \int_y^{y_{max}} \frac{\partial^2}{\partial x^2} g(x, s) \, ds \right]. \tag{3.4}$$

Proof. Using the definition of VLT and denoting $t \doteq \tan \beta$, the function $g(x_v, y_v)$ can be expressed as

$$g(x_v, y_v) = \int_{y_v}^{y_{max}} \frac{f(x_v + (y - y_v) t, y)}{\cos \beta} \, dy + \int_{y_v}^{y_{max}} \frac{f(x_v - (y - y_v) t, y)}{\cos \beta} \, dy.$$

Let $\widehat{g_\lambda}(y_v)$ denote the Fourier transform of $g(x_v, y_v)$ with respect to x_v. Writing it in an expanded form and changing the order of integration, we get

$$\cos \beta \, \widehat{g_\lambda}(y_v) = \int_{y_v}^{y_{max}} \int_{-\infty}^\infty f(x_v + (y - y_v) t, y) \, e^{-i\lambda x_v} dx_v dy$$

$$+ \int_{y_v}^{y_{max}} \int_{-\infty}^\infty f(x_v - (y - y_v) t, y) \, e^{-i\lambda x_v} dx_v dy.$$

By applying the change of variables $X = x_v + (y - y_v)t$ in the first integral and $X = x_v - (y - y_v)t$ in the second, this equation can be rewritten as

$$\cos\beta\,\widehat{g_\lambda}(y_v) = \int_{y_v}^{y_{max}} \int_{-\infty}^{\infty} f(X,y)\,e^{-i\lambda X}e^{i\lambda t(y-y_v)}dX\,dy$$
$$+ \int_{y_v}^{y_{max}} \int_{-\infty}^{\infty} f(X,y)\,e^{-i\lambda X}e^{-i\lambda t(y-y_v)}dX\,dy.$$

It simplifies further by substituting into it the Fourier transform

$$\widehat{f_\lambda}(y) = \int_{-\infty}^{\infty} f(X,y)\,e^{-i\lambda X}dX$$

to obtain

$$\cos\beta\,\widehat{g_\lambda}(y_v) = 2\int_{y_v}^{y_{max}} \widehat{f_\lambda}(y)\cos[\lambda t(y-y_v)]\,dy.$$

Note that here we have achieved the expected diagonalization of the VLT, as $\widehat{g_\lambda}$ depends only on $\widehat{f_\lambda}$ for the same λ.

In order to obtain an explicit formula for $\widehat{f_\lambda}(y)$, we differentiate the above equation with respect to y_v:

$$\cos\beta\,\widehat{g_\lambda}{}'(y_v) = 2\lambda t\int_{y_v}^{y_{max}} \widehat{f_\lambda}(y)\sin[\lambda t(y-y_v)]\,dy - 2\widehat{f_\lambda}(y_v)$$

and combine it with

$$\cos\beta\int_{z}^{y_{max}} \widehat{g_\lambda}(y_v)\,dy_v = \frac{2}{\lambda t}\int_{z}^{y_{max}} \widehat{f_\lambda}(y)\sin[\lambda t(y-z)]\,dy.$$

As a result, we arrive at the following formula:

$$2\,\widehat{f_\lambda}(y_v) = -\cos\beta\,\widehat{g_\lambda}{}'(y_v) + \cos\beta\,(\lambda t)^2\int_{y_v}^{y_{max}} \widehat{g_\lambda}(y)\,dy.$$

Taking the inverse Fourier transform of the last expression yields

$$f(x,y) = -\frac{\cos\beta}{2}\left[\frac{\partial}{\partial y}g(x,y) + \tan^2\beta\int_{y}^{y_{max}} \frac{\partial^2}{\partial x^2}g(x,s)\,ds\right].$$

$\square$

Remark 3.1. Theorem 3.1 is formulated for VLT with the half-opening angle $\beta \in (0, \frac{\pi}{2})$. The case of $\beta = \frac{\pi}{2}$ corresponds to the X-ray transform along the lines with the same fixed angular coefficient. That restricted transform is clearly not injective, thus no inversion can be expected. However, the case $\beta = 0$ does have a meaningful interpretation. From the point of view of the imaging model, it corresponds to the measurements of backscattered radiation. Mathematically, VLT with $\beta = 0$ corresponds to the divergent beam transform multiplied by two. The latter can be inverted trivially using the fundamental theorem of calculus by differentiating the data in the direction of the ray. Formula (3.4) does exactly that when $\beta = 0$.

3.2.2 Conical Radon transform

The approach of diagonalizing the operator through the Fourier transform also works in the case of CRT in slab geometry. The first exact analytic inversion of that transform was derived by Gouia-Zarrad and Ambartsoumian (2014), and we present that result in the following.

Theorem 3.2. *Consider a function* $f \in C_c^\infty(\mathbb{R}^3)$ *supported in* $P = \{(x, y, z) \in \mathbb{R}^3 \mid 0 \le x \le x_{max}, 0 \le y \le y_{max}, 0 \le z \le z_{max}\}$. *For each vertex location* $(x_v, y_v, z_v) \in \mathbb{R}^3$, *denote the CRT of* f *by*

$$g(x_v, y_v, z_v) = C_\beta f(x_v, y_v, z_v). \tag{3.5}$$

An exact inversion of the CRT is given by the formula[2]

$$\widehat{f}_{\lambda,\mu}(z) = k \int_z^{z_{max}} J_0\left[u(z - p)\right] \left[\frac{d^2}{dp^2} + u^2\right]^2 \int_p^{z_{max}} \widehat{g}_{\lambda,\mu}(z_v)\, dz_v\, dp, \tag{3.6}$$

where

$$k = \frac{\cos\beta}{2\pi \tan\beta}, \quad u = \tan\beta\sqrt{\lambda^2 + \mu^2}, \tag{3.7}$$

$\widehat{g}_{\lambda,\mu}(z_v)$ *and* $\widehat{f}_{\lambda,\mu}(z)$ *are, correspondingly, the 2D Fourier transforms of the functions* $g(x_v, y_v, z_v)$ *and* $f(x, y, z)$ *with respect to the first two variables, and* $J_0(z)$ *is the Bessel function of the first kind.*

---

[2] Of course, one needs to take the two-dimensional (2D) inverse Fourier transform of both sides of the equation to recover f.

Proof. Due to the invariance of the family of cones $C(x_v, y_v, z_v)$ with respect to translations along the hyperplane $z = 0$, the conical Radon transform can be expressed as

$$g(x_v, y_v, z_v) = \int_{C(0,0,z_v)} f(x + x_v, y + y_v, z) \, ds.$$

Applying the 2D Fourier transform with respect to variables x_v and y_v yields

$$\widehat{g}_{\lambda,\mu}(z_v) = \int_{C(0,0,z_v)} \widehat{f}_{\lambda,\mu}(z) \, e^{i\lambda x} e^{i\mu y} \, ds. \tag{3.8}$$

Let us denote by $S(z_0, z_v)$ the intersection circle of the cone $C(0, 0, z_v)$ and hyperplane $z = z_0$. We can rewrite the integral with respect to the surface element ds in (3.8) as an integral with respect to dz and dl, where dl is the arc length along the circle $S(z, z_v)$. Namely, with the substitution

$$ds = \frac{dl \, dz}{\cos \beta},$$

equation (3.8) becomes

$$\widehat{g}_{\lambda,\mu}(z_v) = \int_{z_v}^{z_{max}} \widehat{f}_{\lambda,\mu}(z) K_{\lambda,\mu}(z, z_v) \, dz, \tag{3.9}$$

where the kernel $K_{\lambda,\mu}$ is given by the formula

$$K_{\lambda,\mu}(z, z_v) = \frac{1}{\cos \beta} \int_{S(z,z_v)} e^{i(\lambda x + \mu y)} \, dl. \tag{3.10}$$

Using the polar coordinates (r, φ) centered at the location of the vertex (x_v, y_v),

$$x = r \cos \varphi, \quad y = r \sin \varphi, \quad r = \tan \beta \, (z - z_v),$$

the arc length dl can be expressed as

$$dl = r \, d\varphi = \tan \beta \, (z - z_v) \, d\varphi.$$

Substituting the last relation into formula (3.10), we get

$$K_{\lambda,\mu}(z, z_v) = \frac{\tan \beta \, (z - z_v)}{\cos \beta} \int_0^{2\pi} e^{i \tan \beta \, (z - z_v)(\lambda \cos \varphi + \mu \sin \varphi)} \, d\varphi.$$

Now, let ω be the polar angle of vector (λ, μ) in the Fourier domain, i.e. for every $(\lambda, \mu) \neq (0,0)$, the variable ω satisfies the following equalities:

$$\lambda = \sqrt{\lambda^2 + \mu^2} \cos \omega, \quad \mu = \sqrt{\lambda^2 + \mu^2} \sin \omega.$$

Then, the expression of the kernel can be written as

$$K_{\lambda,\mu}(z, z_v) = \frac{\tan \beta \, (z - z_v)}{\cos \beta} \int_0^{2\pi} e^{i \tan \beta \sqrt{\lambda^2 + \mu^2} \, (z - z_v) \cos(\omega - \varphi)} \, d\varphi.$$

Using the integral representation of the Bessel function of the first kind

$$J_0(z) = \frac{1}{2\pi} \int_0^{2\pi} e^{iz \cos \theta} \, d\theta,$$

(e.g. see Abramowitz and Stegun, 1964, p. 360, Section 9.1.21), we can express the kernel in terms of J_0:

$$K_{\lambda,\mu}(z, z_v) = \frac{2\pi \tan \beta}{\cos \beta}(z - z_v) \, J_0 \left[\tan \beta \sqrt{\lambda^2 + \mu^2} \, (z - z_v) \right].$$

As a result, our problem breaks down to the following set of one-dimensional, convolution-type integral equations:

$$\widehat{g}_{\lambda,\mu}(z_v) = k^{-1} \int_{z_v}^{z_{max}} \widehat{f}_{\lambda,\mu}(z) \, (z - z_v) \, J_0 \left[u \, (z - z_v) \right] dz, \qquad (3.11)$$

where k and u are the quantities defined in formula (3.7).

We now show how to solve (3.11) for $\widehat{f}_{\lambda,\mu}(z)$. To that end, we introduce an operator $\mathcal{H}$ defined by

$$\mathcal{H}(F) \doteq \left(\frac{d^2}{dp^2} + u^2 \right) F(p)$$

and apply it to $\int_p^{z_{max}} \widehat{g}_{\lambda,\mu}(z_v) \, dz_v$.

First, we differentiate the above integral using the Leibnitz rule and formula (3.11):

$$k \frac{d^2}{dp^2} \left[\int_p^{z_{max}} \widehat{g}_{\lambda,\mu}(z_v) \, dz_v \right] \qquad (3.12)$$

$$= \int_p^{z_{max}} \widehat{f}_{\lambda,\mu}(z) \, J_0[u(z - p)] \, dz - u \int_p^{z_{max}} \widehat{f}_{\lambda,\mu}(z)(z - p) \, J_1 \left[u(z - p) \right] dz.$$

Here, we took advantage of the fact that $J_0'(z) = -J_1(z)$ (see Abramowitz and Stegun, 1964, p. 361).

Then, using (3.11) and changing the order of integration, we obtain

$$k \int_p^{z_{max}} \widehat{g}_{\lambda,\mu}(z_v) \, dz_v = \int_p^{z_{max}} \widehat{f}_{\lambda,\mu}(z) \int_p^z (z - z_v) \, J_0[u(z - z_v)] \, dz_v \, dz.$$

Applying the integral identity for the Bessel functions $\int_0^p v J_0(v) \, dv = p J_1(p)$ (e.g. see Abramowitz and Stegun, 1964, p. 361, Section 9.1.30), we can rewrite the above equation as follows:

$$k \int_p^{z_{max}} \widehat{g}_{\lambda,\mu}(z_v) \, dz_v = \frac{1}{u} \int_p^{z_{max}} \widehat{f}_{\lambda,\mu}(z)(z - p) \, J_1[u(z - p)] \, dz. \qquad (3.13)$$

Combining equations (3.12) and (3.13), we get

$$k \, \mathcal{H} \left[\int_p^{z_{max}} \widehat{g}_{\lambda,\mu}(z_v) \, dz_v \right] = \int_p^{z_{max}} \widehat{f}_{\lambda,\mu}(z) \, J_0 \left[u(z - p) \right] dz. \qquad (3.14)$$

Note that (3.14) is still a one-dimensional, convolution-type integral equation but with a simpler kernel than in (3.11). We solve it by applying $\mathcal{H}$ again and integrating the result with a Bessel function weight. Namely,

$$k \, \mathcal{H}^2 \left[\int_p^{z_{max}} \widehat{g}_{\lambda,\mu}(z_v) \, dz_v \right]$$
$$= -\widehat{f}_{\lambda,\mu}'(p) + u^2 \int_p^{z_{max}} \widehat{f}_{\lambda,\mu}(z) \left\{ J_0 \left[u(z - p) \right] + J_0'' \left[u(z - p) \right] \right\} dz.$$

Then, we take a weighted integral of both sides of the above equation:

$$k \int_t^{z_{max}} J_0 \left[u(t - p) \right] \mathcal{H}^2 \left[\int_p^{z_{max}} \widehat{g}_{\lambda,\mu}(z_v) dz_v \right] dp$$
$$= - \int_t^{z_{max}} J_0 \left[u(t - p) \right] \widehat{f}_{\lambda,\mu}'(p) dp$$
$$+ u^2 \int_t^{z_{max}} J_0 \left[u(t - p) \right] \int_p^{z_{max}} \widehat{f}_{\lambda,\mu}(z) \left\{ J_0 \left[u(z - p) \right] \right.$$
$$\left. + J_0'' \left[u(z - p) \right] \right\} dz \, dp.$$

The right-hand side (RHS) of the last equation is exactly $\widehat{f}_{\lambda,\mu}(t)$.

To prove that, let us simplify the first term on the RHS by using integration by parts:

$$- \int_t^{z_{max}} J_0 \left[u(t - p) \right] \widehat{f}_{\lambda,\mu}'(p) \, dp = \widehat{f}_{\lambda,\mu}(t) + u \int_t^{z_{max}} J_1 \left[u(t - p) \right] \widehat{f}_{\lambda,\mu}(p) \, dp,$$

where we used the condition imposed on the support of f to conclude that $\widehat{f}_{\lambda,\mu}(z_{max}) = 0$.

To simplify the second term on the RHS, we interchange the order of integration and then use a recurrence relation of Bessel functions (e.g. see Abramowitz and Stegun, 1964, p. 361, Section 9.1.27)

$$J_0''(x) + J_0(x) = J_1(x)/x$$

and the formula (e.g. see Gradshteyn and Ryzhik, 2007, p. 671)

$$\int_0^z J_1(x) \, J_0(z-x) \, \frac{dx}{x} = J_1(z)$$

to get the following:

$$\int_t^{z_{max}} J_0 \left[u(t-p)\right] \int_p^{z_{max}} \widehat{f}_{\lambda,\mu}(z) \left\{ J_0 \left[u(z-p)\right] + J_0'' \left[u(z-p)\right] \right\} dz \, dp$$

$$= \int_t^{z_{max}} \widehat{f}_{\lambda,\mu}(z) \int_t^z J_0 \left[u(t-p)\right] \left\{ J_0 \left[u(z-p)\right] + J_0'' \left[u(z-p)\right] \right\} dp \, dz$$

$$= \frac{1}{u} \int_t^{z_{max}} \widehat{f}_{\lambda,\mu}(z) \int_0^{u(z-t)} J_0 \left[w - u(z-t)\right] \frac{J_1(w)}{w} \, dw \, dz$$

$$= \frac{1}{u} \int_t^{z_{max}} \widehat{f}_{\lambda,\mu}(z) \, J_1 \left[u(z-t)\right] \, dz = -\frac{1}{u} \int_t^{z_{max}} \widehat{f}_{\lambda,\mu}(z) \, J_1 \left[u(t-z)\right] \, dz.$$

Combining all the terms yields the desired result. Thus, we showed that for $(\lambda, \mu) \neq (0,0)$

$$\widehat{f}_{\lambda,\mu}(t) = \int_t^{z_{max}} J_0 \left[u(t-p)\right] \mathcal{H}^2 \left[\int_p^{z_{max}} \widehat{g}_{\lambda,\mu}(z_v) \, dz_v \right] dp.$$

Since f is compactly supported, its Fourier transform at $(\lambda, \mu) = (0,0)$ is uniquely defined by continuity. $\qquad \square$

Remark 3.2. A similar result holds for the generalization of CRT to $\mathbb{R}^d$ for arbitrary $d > 3$ (Gouia-Zarrad, 2014).

3.2.3 Numerical simulations

To demonstrate the performance of inversion formula (3.4) in practice, we present in the following a couple of simple numerical simulations (see Figure 3.2). Both experiments use as original images 256×256 discretized versions of functions supported inside the square $\Omega = [-1,1] \times [-1,1]$. In the first case, the image function is infinitely smooth, whereas the image in the second is piecewise constant.

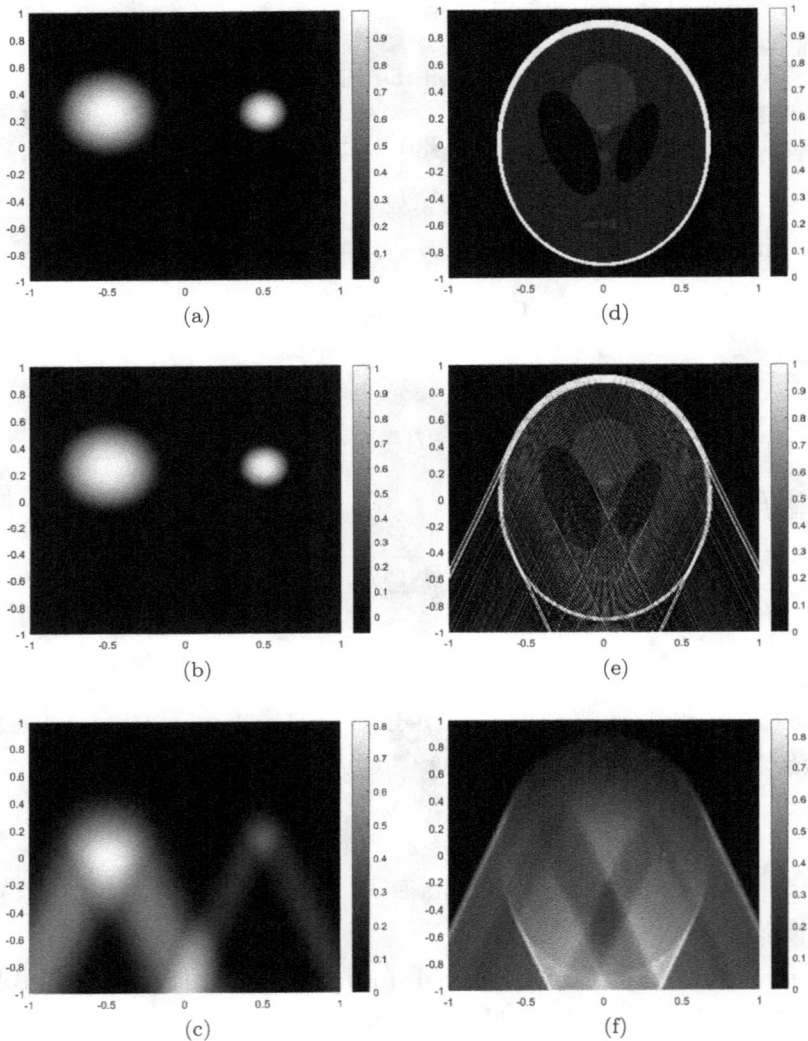

Fig. 3.2   Numerical simulations of VLT in 2D and its inversion using formula (3.4): (a), (d) are the phantoms; (b), (e) are their reconstructions; (c), (f) are the corresponding "sinograms" (VLT data). The colormap in figure (e) has been manually set at $[0, 1]$, while the range of the values of the image is $[-2.46, 4.91]$. Therefore, in (e), all pixels with values above 1 are assigned white color, and those with values below 0 are black. The colormaps in all other figures are based on the true image values. *Implementation of these numerical simulations is courtesy of Souvik Roy.*

Consider the mollifier-like function

$$
f_r(x, y) = \begin{cases} \exp\left[\dfrac{-(x^2 + y^2)}{r^2 - (x^2 + y^2)}\right], & \text{if } x^2 + y^2 < r^2, \\ 0, & \text{otherwise,} \end{cases}
$$

and denote by $f_r^{\mathbf{c}} \doteq f_r(x - c_1, y - c_2)$ the translation of f_r by a vector $\mathbf{c} = (c_1, c_2)$. The phantom used in the first experiment is a sum of two such functions, namely

$$
P_1(x, y) = f_{0.25}^{(0.5, 0.25)}(x, y) + f_{0.4}^{(-0.5, 0.25)}(x, y).
$$

The phantom in the second experiment is a modified Shepp–Logan phantom (MATLAB, 2020).

The VLT data was computed numerically from discretized versions of the function f using the trapezoidal rule. In both experiments, we used a fixed angle $\beta = \pi/8$. The differentiation operators in the inversion formula (3.4) were approximated by the standard first-order forward difference and second-order central difference formulas. The integration in the inversion formula was also done using the trapezoidal rule.

The reconstructions of both phantoms have artifacts (albeit barely visible in the case of the smooth phantom), extending along rays parallel to the direction of the branches of the V-lines. Such artifacts are typical in numerical inversions of this type of transforms (e.g. see Florescu *et al.*, 2011; Gouia-Zarrad and Ambartsoumian, 2014; Zhao *et al.*, 2014; Sherson, 2015; Ambartsoumian and Latifi, 2019, 2021; Walker and O'Sullivan, 2019, 2021). They are especially emphasized at the locations of the vertices of V-lines, whose rays tangentially touch the singularities of f (see Section 3.4 for more discussion about this topic). Various authors have developed efficient strategies to reduce the strengths of such artifacts (e.g. see Felea and Quinto, 2011; Frikel and Quinto, 2013), but the implementation of those techniques and their justification is beyond the scope of this book. For more information about artifacts in numerical inversions of Radon-type transforms, we refer the reader to Felea and Quinto (2011), Frikel and Quinto (2013), Nguyen (2015), Borg *et al.* (2018), and the references therein.

3.3   Cone Differentiation and Integration[3]

In this section, we present another analytic inversion technique for the VLT in slab geometry, which was developed by Ambartsoumian and Latifi (2019). It has an intuitive geometric interpretation, which paves the way for generalizations to the case of weighted VLT in $\mathbb{R}^2$, as well as transforms integrating over polyhedral cones in $\mathbb{R}^n$. In addition to producing exact inversion formulas, the new method also leads to interesting range descriptions and support theorems for these transforms.

The underlying idea of this approach, in a nutshell, can be described as follows. Use the VLT data to obtain a weighted average of the image function on a compact set (a polygon or a polyhedron), and then, take the limit of that quantity (when the size of the set tends to zero) to recover the function. The first step of that process can be accomplished relatively easily due to the presence of "vertices" in the trajectories (surfaces) of integration, which distinguishes the VLT and CRT from the conventional generalized Radon transforms integrating over smooth surfaces.

3.3.1   *Notations and assumptions*

Let us consider weighted VLTs, integrating an image function f along V-lines with a fixed axis of symmetry in the direction of the unit vector $\boldsymbol{\alpha} = (\alpha_x, \alpha_y)$, a constant opening angle 2β, and rays parallel to given unit vectors $\mathbf{u}$ and $\mathbf{v}$ (see Figure 3.3(a)). Since all of the above parameters are fixed, for brevity in the rest of this section, we use the notation

$$\mathcal{B}^{\mathbf{c}}f\,(x,y) \doteq \mathcal{B}^{\mathbf{c}}f\,(\mathbf{r},\Gamma), \tag{3.15}$$

where $\mathbf{r} = (x,y)$, $\Gamma = [\mathbf{u},\mathbf{v}]$ and $\mathbf{c} = (c_u, c_v)$, $c_u \neq 0$, $c_v > 0$. Furthermore, we use two other notations for the special cases of *ordinary* and *signed* VLTs:

$$\mathcal{B}f(x,y) \doteq \mathcal{B}^{(1,1)}f(x,y), \quad \text{and} \quad \mathcal{B}^{-}f(x,y) \doteq \mathcal{B}^{(-1,1)}f(x,y). \tag{3.16}$$

In the definitions above, we assume that $f \in L^1(\mathbb{R}^2)$ and the transformations are defined for almost every $(x,y) \in \mathbb{R}^2$ (for more details, see Section A.3 in the Appendix). With some additional regularity assumption on f (e.g. continuity and compact support), the transformations will be defined at every $(x,y) \in \mathbb{R}^2$.

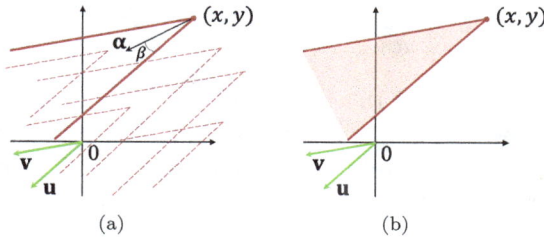

Fig. 3.3 (a) The V-lines in $\mathbb{R}^2$ generated by unit vectors $\mathbf{u}, \mathbf{v}$ and parametrized by the coordinates (x, y) of their vertices. (b) A cone with a vertex at (x, y) generated by unit vectors $\mathbf{u}, \mathbf{v}$. The cone corresponding to $(x, y) = (0, 0)$ is used in this section as the negative cone defining a partial order on $\mathbb{R}^3$.

We use vectors $\mathbf{u}$ and $\mathbf{v}$ as generators of the *negative cone* (see Figure 3.3(b)), which defines a partial order structure on $\mathbb{R}^2$ (see Section A.1 in the Appendix).

In the following sections, we present inversion formulas for the aforementioned transforms and describe some important properties. We do so by using the techniques of cone differentiation and integration developed in Section A.1 of the Appendix.

3.3.2 *Inversion of the ordinary VLT*

Using the moving sections lemma (see Section A.1 in the Appendix), it is easy to prove (see the following) that with the VLT of f, one can generate its integrals over the corresponding negative cones in $\mathbb{R}^2$.

Theorem 3.3. *Let $F(x, y)$ be the integral of $f \in L^1(\mathbb{R}^2)$ over the negative cone at (x, y). Then,*

$$F(x, y) = \sin\beta \int_0^\infty \mathcal{B}f(x + t\alpha_x, y + t\alpha_y)\, dt. \qquad (3.17)$$

Applying the cone differentiation theorem, we immediately obtain an inversion formula for the VLT. Namely, since $|\det(\mathbf{u}, \mathbf{v})| = \sin(2\beta)$, we get the following.

Corollary 3.1 (Two inversion formulas). *Let $f \in C_c(\mathbb{R}^2)$. Then,*

$$f(x, y) = \frac{1}{2\cos\beta} D_u D_v \int_0^\infty \mathcal{B}f(x + t\alpha_x, y + t\alpha_y)\, dt. \qquad (3.18)$$

If it is only known that $f \in L^1(\mathbb{R}^2)$, then we have the following inversion formula for a.e. (x, y):

$$f(x, y) = \lim_{t \to 0} A_t(x, y), \qquad (3.19)$$

where $A_t(x, y)$ is defined as in Theorem A.13 and can be computed as a linear combination of values of F at four points.

Note that in formula (3.18), the directional derivatives are taken in the direction of the generators of the negative cone, while the cone differentiation theorem uses the generators of the positive cone. But since we have a pair of such derivatives, the algebraic signs appearing due to this difference cancel each other.

Remark 3.3. Formula (3.18) is equivalent to (3.4). It has also been obtained by Katsevich and Krylov (2013) and Sherson (2015) using other techniques.

Proof of Theorem 3.3. Let $g = \mathcal{B}f$ and $\gamma(t)$ be the parametric equation of the ray starting at (x, y) and moving in the direction of $\boldsymbol{\alpha} = (\alpha_x, \alpha_y)$, i.e. $\gamma(t) = (x, y) + t\boldsymbol{\alpha}$. This ray divides the region enclosed by the V-line into two parts: A and B (see Figure 3.4).

We apply the moving sections lemma to get the integral of f over these regions. In particular, if we use as moving sections $A_t \doteq V + t\boldsymbol{\alpha}$, then we obtain

$$\int_A f d\mu = \sin\beta \int_0^\infty \int_{A_t} f \, d\mu_V \, dt.$$

Similarly, for $B_t \doteq U + t\boldsymbol{\alpha}$, we get

$$\int_B f d\mu = \sin\beta \int_0^\infty \int_{B_t} f \, d\mu_U \, dt.$$

Adding these two equations, we get the formula

$$F(x, y) = \int_{A \cup B} f d\mu = \sin\beta \int_0^\infty \left(\int_{A_t} f d\mu_U + \int_{B_t} f \, d\mu_V \right) dt$$

$$= \sin\beta \int_0^\infty g(x + t\alpha_x, y + t\alpha_y) \, dt. \qquad \square$$

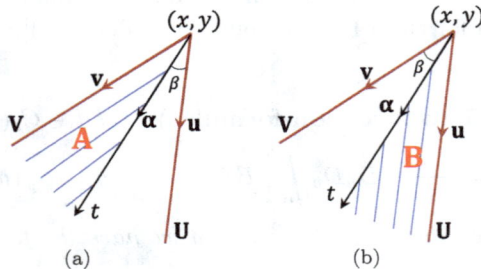

Fig. 3.4   The regions A and B with moving sections.

3.3.3 *Inversion of the weighted VLT*

Consider the weighted V-line transform $\mathcal{B}^c f$ defined by equation (3.15) with constants $c_u \neq 0$ and $c_v > 0$. Let us express the fixed opening angle 2β of the V-line of integration as a sum of two directed angles: $2\beta = \beta_1 + \beta_2 < \pi$ so that

$$\frac{\sin \beta_1}{\sin \beta_2} = \frac{c_v}{c_u} \tag{3.20}$$

(see Figure 3.5). The above relation uniquely defines the angles $\beta_1 \in (0, \pi)$ and β_2 so that $|\beta_2| \in (0, \pi)$. For example, one can use the following identity:

$$\cot \beta_1 = \frac{c_u}{c_v \sin(2\beta)} + \cot(2\beta),$$

and the fact that $\cot x$ is one-to-one on $(0, \pi)$.

Now, let $\tilde{\boldsymbol{\alpha}} = (\tilde{\alpha}_x, \tilde{\alpha}_y)$ be the unique unit vector starting from the vertex of the V-line and satisfying the properties

$$\begin{aligned} \text{angle}\,(\mathbf{v}, \tilde{\boldsymbol{\alpha}}) &= \beta_1, \\ \text{angle}\,(\tilde{\boldsymbol{\alpha}}, \mathbf{u}) &= \beta_2. \end{aligned} \tag{3.21}$$

The vector $\tilde{\boldsymbol{\alpha}}$ can also be expressed as a linear combination of the unit vectors $\mathbf{u}$ and $\mathbf{v}$ as follows:

$$\tilde{\boldsymbol{\alpha}} = \frac{c_u \mathbf{v} + c_v \mathbf{u}}{\|c_u \mathbf{v} + c_v \mathbf{u}\|}. \tag{3.22}$$

To verify the last relation, we use the cross products:

$$\sin \beta_1 = \|\mathbf{v} \times \tilde{\boldsymbol{\alpha}}\| = \frac{\|\mathbf{v} \times (c_u \mathbf{v} + c_v \mathbf{u})\|}{\|c_u \mathbf{v} + c_v \mathbf{u}\|} = \frac{c_v \sin(2\beta)}{\|c_u \mathbf{v} + c_v \mathbf{u}\|},$$

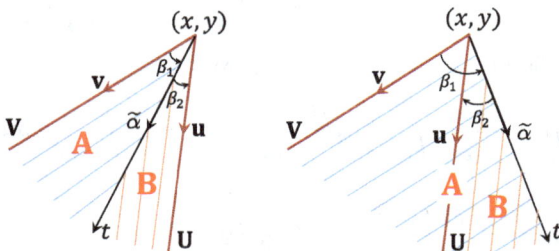

Fig. 3.5   The new direction of integration $\tilde{\boldsymbol{\alpha}}$ and the moving sections. The sketch on the left depicts the setup when $c_u > 0$, while the one on the right corresponds to $c_u < 0$.

$$|\sin \beta_2| = \|\widetilde{\boldsymbol{\alpha}} \times \mathbf{u}\| = \frac{\|(c_u \mathbf{v} + c_v \mathbf{u}) \times \mathbf{u}\|}{\|c_u \mathbf{v} + c_v \mathbf{u}\|} = \frac{|c_u| \sin (2\beta)}{\|c_u \mathbf{v} + c_v \mathbf{u}\|}.$$

Note that formula (3.22) implies that sgn $(\sin \beta_2)$ = sgn (c_u). As a result,

$$\sin \beta_2 = \frac{c_u \sin (2\beta)}{\|c_u \mathbf{v} + c_v \mathbf{u}\|},$$

and formula (3.20) follows immediately.

Just as in the case of the ordinary VLT, to invert the weighted VLT, we express $F(x,y)$ through $\mathcal{B}^c f$ and apply the cone differentiation theorem.

Theorem 3.4. *Let $F(x,y)$ be the integral of $f \in L^1(\mathbb{R}^2)$ over the negative cone at (x,y). If c_v, c_u, β_1, β_2, and $\widetilde{\alpha}$ are defined as above, then*

$$F(x,y) = \frac{\sin \beta_1}{c_v} \int_0^\infty \mathcal{B}^c f(x + \widetilde{\alpha}_x t, y + \widetilde{\alpha}_y t) \, dt. \tag{3.23}$$

Proof. We follow the same steps as in the proof of the nonweighted version, but this time, we divide the integration region using a new direction $\widetilde{\alpha}$ defined in the statement of the theorem.

Applying the moving sections lemma, we obtain

$$\int_A f \, d\mu = \sin \beta_1 \int_0^\infty \int_{V+t\widetilde{\alpha}} f \, d\mu_V \, dt,$$

$$\int_B f \, d\mu = \sin \beta_2 \int_0^\infty \int_{U+t\widetilde{\alpha}} f \, d\mu_U \, dt.$$

Note that if $c_u > 0$, then $F(x,y) = \int_{A \cup B} f \, d\mu$, while in the case of $c_u < 0$, we have $F(x,y) = \int_{A \setminus B} f \, d\mu$.

In both cases, we get

$$F(x,y) = \int_0^\infty \left(\sin \beta_1 \int_{V+t\widetilde{\alpha}} f \, d\mu_V + \sin \beta_2 \int_{U+t\widetilde{\alpha}} f \, d\mu_U \right) dt.$$

Using the relation $c_v/c_u = \sin \beta_1 / \sin \beta_2$, we have

$$F(x,y) = \frac{\sin \beta_1}{c_v} \int_0^\infty \left(c_v \int_{V+t\widetilde{\alpha}} f \, d\mu_V + c_u \int_{U+t\widetilde{\alpha}} f \, d\mu_U \right) dt$$

$$= \frac{\sin \beta_1}{c_v} \int_0^\infty \mathcal{B}^c f(x + \widetilde{\alpha}_x t, y + \widetilde{\alpha}_y t) \, dt,$$

which finishes the proof. $\qquad\qquad\square$

In the special case of the signed VLT (i.e. $c_v = -c_u = 1$), we have $\tilde{\alpha} = (\mathbf{u} - \mathbf{v})/\|\mathbf{u} - \mathbf{v}\|$, which is the unit vector perpendicular to the cone direction α. As a result, we get the following.

Theorem 3.5. *Let $F(x, y)$ be the integral of $f \in L^1(\mathbb{R}^2)$ over the negative cone at (x, y) and let $\mathcal{B}^- f$ be the signed V-line transform of f. For $\tilde{\alpha} = (\mathbf{u} - \mathbf{v})/\|\mathbf{u} - \mathbf{v}\|$, we have*

$$F(x, y) = \cos \beta \int_0^\infty \mathcal{B}^- f(x + \tilde{\alpha}_x t, y + \tilde{\alpha}_y t) \, dt. \tag{3.24}$$

Applying the cone differentiation theorem to Theorem 3.4, one can now get an inversion formula for the weighted VLT.

Theorem 3.6 (Inversion formula). *Let $\mathcal{B}^c f$ be the weighted V-line transform of $f \in C_c(\mathbb{R}^2)$, with arbitrary nonzero weights, and*

$$\tilde{\alpha} = \frac{c_u \mathbf{v} + c_v \mathbf{u}}{\|c_u \mathbf{v} + c_v \mathbf{u}\|}.$$

Then,

$$f(x, y) = \frac{1}{\|c_u \mathbf{v} + c_v \mathbf{u}\|} D_u D_v \int_0^\infty \mathcal{B}^c f(x + \tilde{\alpha}_x t, y + \tilde{\alpha}_y t) \, dt. \tag{3.25}$$

The coefficient in the above formula can be expressed as

$$\frac{1}{\|c_u \mathbf{v} + c_v \mathbf{u}\|} = \frac{\sin \beta_1}{c_v \sin (2\beta)}. \tag{3.26}$$

Remark 3.4. Just as in the case of the ordinary VLT, if it is only known that $f \in L^1(\mathbb{R}^2)$, for $\mathcal{B}^c$ too, we have the following inversion formula for a.e. (x, y):

$$f(x, y) = \lim_{t \to 0} A_t(x, y),$$

where $A_t(x, y)$ is defined as in Theorem A.13 and can be computed as a linear combination of the values of F at four points.

Remark 3.5. Note that formula (3.25) holds even if one of the weight coefficients, c_u or c_v, is equal to zero. In that case, the inversion of VLT reduces to a trivial application of the FTC. The weighted VLT in such a setup is essentially (up to a constant multiple) the same as the ordinary VLT that uses V-lines with an opening angle $2\beta = 0$, i.e. with coinciding branches.

We finish this section with a statement about the "interior problem" of inverting VLT with some boundary information.

Assume that $f \in C_c(\mathbb{R}^2)$ and its values are known on the boundary of some bounded, open, convex set Ω. Then, one can recover f inside Ω using its weighted VLT data restricted to V-lines with vertices inside Ω. Namely, we have the following.

Theorem 3.7 (Inversion of a restricted VLT). *Consider a bounded, open, convex set Ω in $\mathbb{R}^2$ and let $g = \mathcal{B}^c f$ be the weighted V-line transform of $f \in C_c(\mathbb{R}^2)$. For each point $\boldsymbol{p} = (x, y) \in \Omega$, let m_p be the distance to the boundary along the vector $\widetilde{\boldsymbol{\alpha}}$, i.e. $(x_0, y_0) \doteq \boldsymbol{p} + m_p \widetilde{\boldsymbol{\alpha}} \in \partial\Omega$. Then,*

$$f(x, y) = f(x_0, y_0) + \frac{\sin \beta_1}{c_v \sin (2\beta)} D_u D_v \int_0^{m_p} g(\boldsymbol{p} + s\widetilde{\boldsymbol{\alpha}}) \, ds. \qquad (3.27)$$

The proof follows trivially from the previous result and the additivity of integration along line segments.

Remark 3.6. In the case of the signed VLT (i.e. $c_v = -c_u = 1$), a similar inversion formula was obtained by Katsevich and Krylov (2013) using other techniques.

3.3.4 *Range description*

One of the most important tasks related to any generalized Radon transform is the description of its range (e.g. see Aguilar and Kuchment, 1995; Aguilar *et al.*, 1996; Novikov, 2002b; Ambartsoumian and Kuchment, 2006; Finch and Rakesh, 2006; Agranovsky *et al.*, 2007, 2009; Agranovsky and Nguyen, 2010). In addition to a purely theoretical interest, the range characterization has valuable practical implications for the associated imaging modalities. Typically, the range of a GRT has infinite codimension in standard function spaces, i.e. a function in its range must satisfy infinitely many conditions. In the imaging community, the latter are often called *consistency conditions*. When a GRT represents measured data of a tomographic device, these conditions can be used to suppress noise in measurements, complete missing data, identify hardware imperfections, etc. (e.g. see Natterer, 1983; Hertle, 1988; Lvin, 1994; Solmon, 1995; Ponomaryov, 1995; Clarkson, 1999; Mennessier *et al.*, 1999; Anastasio *et al.*, 2001; Patch, 2004).

Range characterization for VLT and CRT with vertices inside the image domain is a scarcely studied subject. To the best of our knowledge, Katsevich and Krylov (2013), Ambartsoumian and Latifi (2019), and Baines

(2021) are the only ones who have investigated the problem in a formally determined setup (i.e. for the VLT with two or CRT with three degrees of freedom). In this section, we present the results of Ambartsoumian and Latifi (2019). The work by Katsevich and Krylov (2013) is discussed in Section 4.2. The main ideas of Baines (2021) are outlined in Section 3.5.

We start with providing the necessary and sufficient conditions for a function F to be a cone integral of another function $f \geq 0$ with respect to a given order structure in $\mathbb{R}^n$. In other words, we answer the question: For which F is there an $f \geq 0$ such that $F(\mathbf{x}) = \int_{\mathbf{y} \leq \mathbf{x}} f(\mathbf{y}) \, d\mu$?

Our approach is motivated by the corresponding result for $\mathbb{R}^1$ stated in Remark A.10. We use the Radon–Nikodym theorem to get the desired description of F. For an appropriate F, we construct a corresponding measure ν, for which Theorem A.12 implies the existence of its Radon–Nikodym derivative f.

For $\mathbf{x} \in \mathbb{R}^n, c_i \in \mathbb{R}^+$, let $P(\mathbf{x}, c_1, \dots, c_n)$ be an n-dimensional half-open parallelepiped defined by

$$P(\mathbf{x}, c_1, \dots, c_n) = \left\{ \mathbf{x} + \sum_{i=1}^{n} t_i \mathbf{v}_i \in \mathbb{R}^n; -c_i \leq t_i < c_i \right\}, \qquad (3.28)$$

where $\mathbf{v}_1, \dots, \mathbf{v}_n$ are the basis vectors defining the order structure in $\mathbb{R}^n$. These parallelepipeds are the analogs of the intervals in $\mathbb{R}^1$.

For a given function $F : \mathbb{R}^n \to \mathbb{R}$, we define a set function ν_0 on the ring of subsets generated by these parallelepipeds by

$$\nu_0(P(\mathbf{x}, c_1, \dots, c_n)) \doteq \sum_{\sigma \in \{-1,1\}^n} \mathrm{sgn}(\sigma_1 \dots \sigma_n) F(\mathbf{x} + \sigma_1 c_1 \mathbf{v}_1 + \dots + \sigma_n c_n \mathbf{v}_n)$$

and extending ν_0 to the ring using the outer measure induced by ν_0.

Using an analogy with absolute continuity on the real line, we define absolute continuity of F as follows.

Definition 3.4. Let F be a function on $\mathbb{R}^n$ with a given order structure. Then, we say that F is *absolutely continuous* if and only if for every $\epsilon > 0$, there exists $\delta > 0$ such that for any finite collection of disjoint parallelepipeds $\{P_i\}_{i=1}^{m}$, $\sum_{i=1}^{m} \mu(P_i) < \delta$ implies $\sum_{i=1}^{m} |\nu_0(P_i)| < \epsilon$.

Let us call

$$\alpha \doteq \frac{\mathbf{v}_1 + \dots + \mathbf{v}_n}{\|\mathbf{v}_1 + \dots + \mathbf{v}_n\|}$$

the direction of the negative cone.

Definition 3.5. We say that a function F on $\mathbb{R}^n$ is *P-cumulative* with respect to the given order structure if

- $\nu_0(P) \geq 0$ for every parallelepiped P (nondecreasing condition);
- $\lim\limits_{t \to \infty} F(t\boldsymbol{\alpha}) = 0$ and $\lim\limits_{t \to -\infty} F(t\boldsymbol{\alpha}) < \infty$.

The second condition may appear reversed for readers familiar with cumulative probability distributions. It is simply due to the fact that we use integrals over negative cones and $\boldsymbol{\alpha}$ points in the negative direction.

Remark 3.7. Note that if $f \geq 0$ is integrable, then F defined by formula (A.2) will be P-cumulative with respect to the underlying order structure.

If F is P-cumulative, then ν_0 is a pre-measure on the ring of subsets generated by the parallelepipeds. By applying the Caratheodory extension theorem, we can extend ν_0 to a measure ν on $\mathbb{R}^n$ (the domain of ν is the σ-algebra of Lebesgue measurable sets). The details of this construction can be found in many measure theory texts (see for example Athreya and Lahiri, 2006).

We observe that for $F(\mathbf{x}) = \int_{\mathbf{y} \leq \mathbf{x}} f(\mathbf{y})\, d\mu$, if we send $\mathbf{x}$ to infinity in the direction opposite to $\boldsymbol{\alpha}$, then in the limit, we get the integral of f over $\mathbb{R}^n$. In other words,

$$\lim_{t \to -\infty} F(t\boldsymbol{\alpha}) = \int_{\mathbb{R}^n} f\, d\mu.$$

Hence, for nonnegative f, if $\lim\limits_{t \to -\infty} F(t\boldsymbol{\alpha}) < \infty$, we have $f \in L^1(\mathbb{R}^n)$.

The following is a consequence of the Radon–Nikodym theorem (see Theorem A.12 in the Appendix):

Lemma 3.1. *Let F be a P-cumulative function on $\mathbb{R}^n$. Then, F is absolutely continuous if and only if there exists a nonnegative function f such that*

$$F(\boldsymbol{x}) = \int_{y \leq x} f\, d\mu.$$

Proof. Assume $f \geq 0$. Then (as we discussed above), f is integrable and therefore locally integrable. This implies that f is the density of some absolutely continuous measure $\lambda \ll \mu$, i.e.

$$\lambda(E) = \int_E f\, d\mu.$$

Note that since $f \in L^1(\mathbb{R}^2)$, it follows that λ is a finite measure; hence, one can use the "$\epsilon - \delta$" definition of absolute continuity (see Section A.2 in the Appendix). Hence, for any $\epsilon > 0$, there exists a $\delta > 0$ such that $\mu(E) < \delta$ implies $\lambda(E) < \epsilon$.

Combining the above relation with

$$\sum_{i=1}^{m} |\nu_0(P_i)| = \sum_{i=1}^{m} \int_{P_i} f \, d\mu = \int_{\cup P_i} f \, d\mu,$$

we establish that F is absolutely continuous in the sense of Definition 3.4.

The other direction of the statement is an implication of the Radon–Nikodym theorem. Here, we use the fact that the constructed measure ν is absolutely continuous with respect to μ if F is absolutely continuous in the sense of Definition 3.4 (see Section A.2 in the Appendix for more details). □

Corollary 3.2. *If f is related to F, as described in the previous theorem, then for any measurable set S, we have $\nu(S) = \int_S f \, d\mu$.*

We use the previous lemma to provide a range description for the VLT. But first, let us introduce one more notation and a technical result that will be handy in the proof of the main theorem. For $(x, y) \in \mathbb{R}^2$, let

$$S_l(x, y) \doteq \{(x_1, y_1) \,|\, (x_1, y_1) \leq (x, y)\} \setminus \{(x_1, y_1) \,|\, (x_1, y_1) \leq (x, y) + \alpha l\}$$

be the "broken strip" region of width $l \sin \beta > 0$ (see Figure 3.6).

Lemma 3.2. *For $f \in L^1(\mathbb{R}^2)$ and $(x, y) \in \mathbb{R}^2$, we have*

$$\lim_{\varepsilon \to 0} \frac{1}{\varepsilon} \int_{S_\varepsilon(x,y)} f \, d\mu = \sin \beta \int_{L(x,y)} f \, dl$$

almost everywhere, where $L(x, y)$ is the V-line at (x, y).

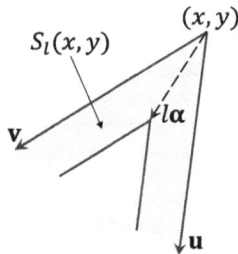

Fig. 3.6   The "broken strip" $S_l(x, y)$.

Proof. Define

$$k(t) \doteq \int_{L((x,y)+t\alpha)} f \, dl.$$

By the moving sections lemma, we have

$$\int_{S_\varepsilon(x,y)} f \, d\mu = \sin \beta \int_0^\varepsilon k(t) dt.$$

Dividing both sides by ε, taking a limit, and applying Theorem A.11 (Lebesgue differentiation theorem), we get

$$\lim_{\varepsilon \to 0} \frac{1}{\varepsilon} \int_{S_\varepsilon(x,y)} f \, d\mu = \sin \beta \lim_{\varepsilon \to 0} \frac{1}{\varepsilon} \int_0^\varepsilon k(t) \, dt = \sin \beta \, k(0) = \sin \beta \int_{L(x,y)} f \, dl,$$

for almost every $(x, y) \in \mathbb{R}^2$. The use of Theorem A.11 is valid since $k(t) \in L^1(\mathbb{R})$ by Fubini's theorem applied to $f \in L^1(\mathbb{R}^2)$. □

Now, we use Lemmas 3.1 and 3.2 to prove the main result of this section.

Theorem 3.8 (Range description). *For a given $g(x, y)$, there exists some nonnegative function $f \in L^1(\mathbb{R}^2)$ such that $g = \mathcal{B}f$ if and only if the function F defined by*

$$F(x, y) = \sin \beta \int_0^\infty g(x + t\alpha_x, y + t\alpha_y) \, dt$$

is P-cumulative and absolutely continuous in the sense of Definition 3.4.

Proof. Let g be the image of a nonnegative function $f \in L^1(\mathbb{R}^2)$ under the VLT. Then, we have

$$F(x, y) = \sin \beta \int_0^\infty \int_{L(x+t\alpha_x, y+t\alpha_y)} f \, dl \, dt = \int_{z \le (x,y)} f \, d\mu$$

and hence by Theorem 3.1, F is absolutely continuous and P-cumulative. For the proof in the other direction, assume that F is absolutely continuous and P-cumulative. By Theorem 3.1, there exists a nonnegative function f such that

$$F(x, y) = \sin \beta \int_0^\infty g(x + t\alpha_x, y + t\alpha_y) \, dt = \int_{\mathbf{z} \le (x,y)} f \, d\mu.$$

At the same time, Lemma 3.2 implies that for almost every $(x, y) \in \mathbb{R}^2$,

$$\lim_{\varepsilon \to 0} \frac{F(x, y) - F(x + \varepsilon \alpha_x, y + \varepsilon \alpha_y)}{\varepsilon} = \sin \beta \int_{L(x,y)} f \, d\mu.$$

On the other hand, we can rewrite the left-hand side of this relation as follows and apply Theorem A.11:

$$\lim_{\epsilon \to 0} \frac{1}{\epsilon} \int_0^\epsilon g(x + t\alpha_x, y + t\alpha_y) \sin \beta \, dt = \sin \beta \, g(x, y)$$

for almost every (x, y). Hence, $g(x, y) = \int_{L(x,y)} f \, d\mu$ almost everywhere. Also, f is integrable because

$$\int_{\mathbb{R}^2} f \, d\mu = \lim_{t \to -\infty} F(t\alpha_x, t\alpha_y) = \sin \beta \int_{-\infty}^{+\infty} g(t\alpha_x, t\alpha_y) \, dt < \infty. \qquad \square$$

A similar result with essentially the same proof holds for the weighted VLT.

Theorem 3.9 (Range of the weighted VLT). *For a given $g(x, y)$, there exists some nonnegative function $f \in L^1(\mathbb{R}^2)$ such that $g = \mathcal{B}^c f$ if and only if the function F, as defined in Theorem 3.4, is P-cumulative and absolutely continuous.*

3.3.5 Support theorems for the VLT

Theorem 3.10. *Let $\mathcal{B}^c f$ be the weighted V-line transform of $f \in L^1(\mathbb{R}^2)$ and F be defined as in formula (3.23). If F is constant on some $S \subset \mathbb{R}^2$ with a nonempty interior S^0, then $f = 0$ in S^0.*

Proof. Let x be an interior point of S. Then, we can find t such that all parallelograms of size smaller than t centered at x will lie inside S. Since F has the same values at the corners of these parallelograms, by definition, we have $A_t(x) = 0$. Hence, $f(x) = \lim_{t \to 0} A_t(x) = 0$ by Theorem A.13. $\qquad \square$

Theorem 3.11. *Let $f \in L^1(\mathbb{R}^2)$ be continuous and $\mathcal{B}^c f = 0$ on some $S \subset \mathbb{R}^2$. Let L be a line parallel to the vector $\tilde{\alpha}$ defined in (3.22) such that $L \cap S \neq \varnothing$. Then, f is constant on each connected component of $L \cap S$.*

In particular, if S is a compact set, $\mathcal{B}^c f = 0$ in S and $f = 0$ on the boundary ∂S of S, then $f \equiv 0$ in S.

Proof. Take two points on a connected component of $L \cap S$, and let Ω be the line interval connecting these two points. Then, Ω is a compact convex set, and by Theorem 3.7, the value of f should be the same at both endpoints.

The second part follows from the fact that for any x in the closed and bounded S, the intersection of the ray $L = \{x + t\alpha \mid t \in \mathbb{R}\}$ and ∂S is nonempty. $\qquad \square$

3.3.6   *Radon transform over polyhedral cones*

Now, we consider a generalization of our results to higher dimensions. As before, we assume that $f \in L^1(\mathbb{R}^n)$, and the transformation is defined at almost every $\mathbf{x} \in \mathbb{R}^n$.

Definition 3.6. The *p-conical Radon transform* T_p maps f into the set of its integrals over the boundaries $\partial C(\mathbf{x})$ of polyhedral cones $C(\mathbf{x})$ generated by fixed unit basis vectors $\mathbf{u}_1, \ldots, \mathbf{u}_n$ starting from $\mathbf{x}$ (see Figure 3.7). Namely,

$$(T_p f)(\mathbf{x}) = \int_{\partial C(\mathbf{x})} f \, dS, \tag{3.29}$$

where dS is the standard $n-1$-dimensional Lebesgue measure on ∂C.

Note that the number of edges (and faces) of the polyhedral cone coincides with the dimension of the underlying space. In this case, there exists a unique unit vector $\boldsymbol{\alpha}$ such that when starting from the vertex $\mathbf{x}$ of the cone, $\boldsymbol{\alpha}$ is pointing inside the cone and has the same angle β with all n faces of the cone.

Let $X_i = \text{span}\langle \mathbf{u}_1, \ldots, \mathbf{u}_{i-1}, \mathbf{u}_{i+1}, \ldots, \mathbf{u}_n \rangle$ denote the hyperplane containing the ith face of the polyhedral cone and define $\mathbf{y}_i$ to be the unit vector in $X_i^\perp$ such that

$$\langle \boldsymbol{\alpha}, \mathbf{y}_i \rangle = \sin \beta, \quad \forall i = 1, \ldots, n.$$

The following theorem is the n-dimensional analogue of Theorem 3.3, as it provides a formula for generating the integral of f over polyhedral cones C from the conical Radon transforms $T_p f$.

Theorem 3.12. *Let* $f \in L^1(\mathbb{R}^n)$ *and* $T_p, \alpha, \mathbf{y}_j$ *be defined as above. Then,*

$$F(\boldsymbol{x}) = \langle \boldsymbol{\alpha}, \boldsymbol{y}_1 \rangle \int_0^\infty (T_p f)(\boldsymbol{x} + t\boldsymbol{\alpha}) \, dt \tag{3.30}$$

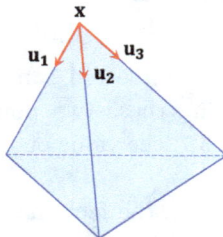

Fig. 3.7   A polyhedral cone in $\mathbb{R}^3$ generated by unit vectors $\mathbf{u}_1$, $\mathbf{u}_2$, and $\mathbf{u}_3$.

is the integral of f over the cone generated by $\mathbf{u}_1, \ldots, \mathbf{u}_n$ with the vertex at $\mathbf{x}$.

Proof. We use a strategy similar to the 2D case by replacing regions A, B in that proof with $\{A_i\}_{i=1}^n$. Let $C(\mathbf{x})$ be the cone at $\mathbf{x}$. Then, $C = \cup_{i=1}^n A_i$, where A_i is the cone made by $\mathbf{u}_1, \ldots, \mathbf{u}_{i-1}, \mathbf{u}_{i+1}, \ldots, \mathbf{u}_n$ and $\boldsymbol{\alpha}$. In other words, we break C into disjoint components using the vector $\boldsymbol{\alpha}$.

Now, to integrate f over C, we write

$$\int_C f \, d\mu = \sum_{i=1}^n \int_{A_i} f \, d\mu.$$

A straightforward application of the moving sections lemma yields

$$\int_{A_i} f \, d\mu = \sin \beta \int_0^\infty \int_{X_i + t\boldsymbol{\alpha}} f \, d\mu_X \, dt.$$

At the same time,

$$\sum_{i=1}^n \int_{X_i + t\boldsymbol{\alpha}} f \, d\mu_X = T_p f(\mathbf{x} + t\boldsymbol{\alpha}),$$

which finishes the proof. $\qquad\qquad\qquad\qquad\qquad\qquad\qquad\qquad\qquad$ $\square$

Corollary 3.3. *One can invert the p-conical Radon transform T_p by using formula (3.30) to generate F from $T_p f$ and then applying Theorem* A.13 *or Theorem* A.14.

Remark 3.8. In analogy with the 2D case, one can consider a weighted p-conical Radon transform where the integration along each face of the polyhedral cone is done with a different constant weight. An inversion procedure for such a transform can be obtained following the approach of the 2D case and the previous corollary.

Remark 3.9. Theorems 3.7, 3.10, and 3.11 can all be generalized to the case of the p-conical Radon transform, with proofs following the corresponding arguments used in the case of the V-line transform.

Remark 3.10. A range description for the p-conical Radon transform may be derived with an appropriate (albeit very tedious) generalization of the techniques used in the case of the VLT.

Remark 3.11. We are not aware of any imaging application for the p-conical Radon transforms. At this point, the study of such transformations is of purely theoretical interest.

3.3.7   *Numerical simulations*

In this section, we show some results of the numerical implementation of VLT inversion algorithms based on cone differentiation. Our simulations have revealed that the reconstructions using formula (3.18) are virtually indistinguishable from those using (3.4) presented in Figure 3.2 of Section 3.2.3. Therefore, in the following, we demonstrate the efficiency of VLT inversion algorithms using formula (3.19) (see Figure 3.9).

The original image is an 800×800 Shepp–Logan phantom from MAT-LAB (2020) (see Figure 3.8(a)). The VLT is computed along V-lines with a half-opening angle $\beta = \arctan(1/2)$ and a horizontal symmetry axis (i.e. $\boldsymbol{\alpha} = (-1,0)$). To compute the one-dimensional ray integrals, we use linear interpolation to evaluate the image values along a given line with a step size of $dx = 0.8$ (in pixels). The VLT data is then generated by adding two ray integrals in two directions (see Figure 3.8(b)).

Our reconstructions are based on formula (3.19), which uses the averages of f over infinitesimal parallelograms with a side length $t = \epsilon$ (see Theorem A.13). In each experiment, we have chosen a specific size for ϵ (measured in pixels). To get the best outcome, one needs to strike an appropriate balance for that size, since large values of ϵ produce a blurry outcome, while small ones lead to more artifacts. The need to refine the choice of ϵ becomes even more important in the presence of noise in the VLT data. We have illustrated the effects of added Gaussian noise in Figures 3.9(c)–(f). To test the effects of data preprocessing, in the last two reconstructions,

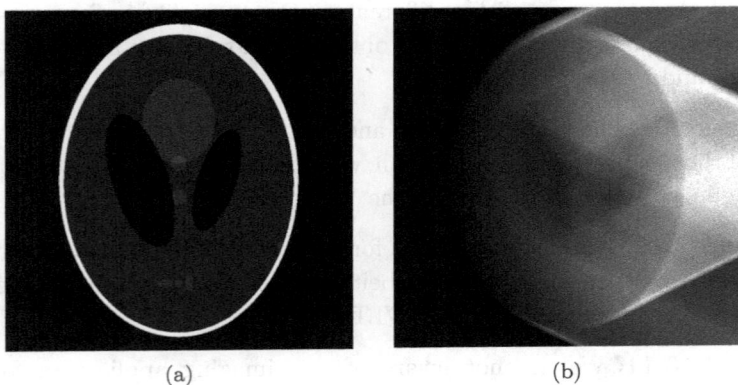

(a)   (b)

Fig. 3.8   (a) Shepp–Logan phantom; (b) the corresponding "sinogram" (VLT data). *Implementation of these numerical simulations is courtesy of Mohammad J. Latifi.*

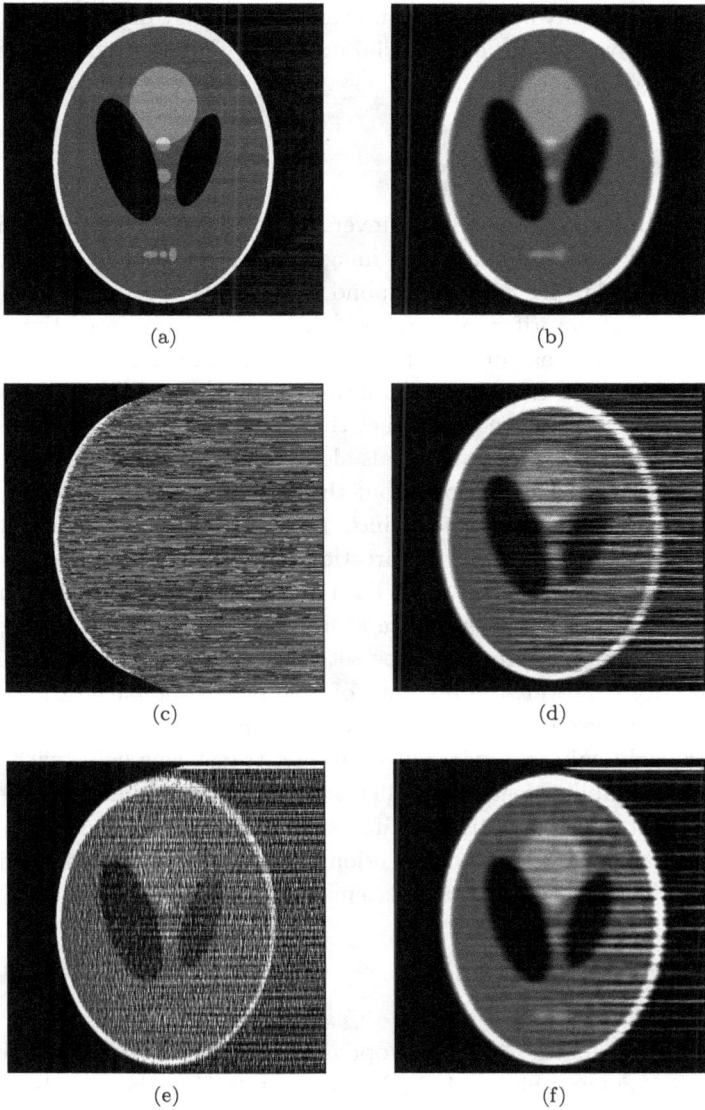

Fig. 3.9 Numerical simulations of VLT inversion using formula (3.19), which averages f over infinitesimal parallelograms with a side length $t = \epsilon$: (a) 0% noise, ϵ=1; (b) 0% noise, ϵ=10; (c) 5% noise, ϵ=1; (d) 5% noise, ϵ=10; (e) 5% noise, ϵ=1, window=10; (f) 5% noise, ϵ=10, window=10. *Implementation of these numerical simulations is courtesy of Mohammad J. Latifi.*

we have applied convolution-type filtering to the noisy data. The filters are distinguished by the size of the "sliding window" measured in pixels (see Figures 3.9(e)–(f)).

3.4   Microlocal Analysis

Numerical simulations of all the inversion formulas for the VLT in slab geometry show decent quality of image reconstruction, despite the fact that the available VLT data correspond to integrals of f (essentially) along two fixed directions (if one "forgets" about the presence of the vertices of V-lines). In the case of the ordinary Radon transform, any attempt to recover the function from its integrals along the lines with two fixed directions will fail. This is due to the fact that a generalized Radon transform, mapping a function f to its integrals along a family of *smooth* curves, preserves only certain information about the wavefront set of f (i.e. the set of locations $\mathbf{x}$ and directions $\boldsymbol{\xi}$ for which f is not smooth). Namely, if f has a singularity at a point $\mathbf{x}$ in the direction $\boldsymbol{\xi} \in \mathbb{S}^1$, then it can be recovered stably from the generalized Radon data only if the latter includes the integration curves passing through $\mathbf{x}$ and normal to $\boldsymbol{\xi}$. Therefore, in the case of only two available directions, the set of "visible" singularities will have measure zero. However, the presence of a vertex in the path of integration completely changes the microlocal properties of the transform.

Microlocal analysis of generalized Radon transforms helps with understanding the stability features of their inversion and explains the presence, nature, and strength of various artifacts in the reconstructions. Before presenting the main results of this section, we give the formal definitions of the central concepts used in the statements.

3.4.1   *The wavefront set*

Definition 3.7. Let $\Omega \subset \mathbb{R}^2$ be an open set and $f \in \mathcal{D}'(\Omega)$ be a distribution (e.g. see Gelfand and Shilov, 1964; Hörmander, 2015). The distribution f is called C^∞ (or *smooth*) on Ω if there exists a function $F(\mathbf{x}) \in C^\infty(\Omega)$ such that

$$(f, \phi) = \int F(\mathbf{x}) \, \varphi(\mathbf{x}) \, d\mathbf{x}$$

for all test functions $\varphi \in \mathcal{D}(\Omega)$ supported in Ω.

Definition 3.8. The *singular support of f*, denoted by sing supp(f), is the complement of the union of all open sets on which f is C^∞.

In other words, a point $\mathbf{x} \in \Omega$ is *not* in the sing supp(f) if f is smooth in some neighborhood of $\mathbf{x}$.

In addition to locating the singularity of a function (or a distribution) in the space, it is also helpful to identify the direction of the singularity. For example, the characteristic function of the unit square, $\chi_{[0,1] \times [0,1]}(\mathbf{x})$, is singular at every point of the set $\{(x_1, 0) : 0 < x_1 < 1\}$, but it is smooth in the direction of the vector $(1, 0)$ at each point of that set. To formalize the notion of the direction of singularity, let us recall a result from Fourier analysis establishing a relation between the smoothness of a function and the decay at infinity of its Fourier transform.

Definition 3.9. A function $F : \mathbb{R}^2 \to \mathbb{C}$ is called *rapidly decaying at infinity* if for every $N \geq 0$, there exists a constant C_N such that for all $\mathbf{x} \in \mathbb{R}^2$

$$|F(\mathbf{x})| \leq C_N \left(1 + ||\mathbf{x}||\right)^{-N}.$$

Theorem 3.13 (Rudin (1973)). *Consider a function $F \in L^2(\Omega)$ compactly supported inside Ω. Then, $F \in C_c^\infty(\Omega)$ if and only if its Fourier transform $\widehat{F}$ is rapidly decaying at infinity.*

We can use the above result to give an alternative description of the singular support of a distribution f. Namely, assume that $\mathbf{x}_0 \notin$ sing supp(f). Then, there exists a neighborhood U of $\mathbf{x}_0$ such that f is smooth inside U. Let us choose a test function $\varphi \in C_c^\infty(U)$ such that $\varphi(\mathbf{x}_0) \neq 0$. The product φf is a smooth function compactly supported inside U; therefore, $\widehat{\varphi f}$ will be rapidly decaying at infinity, i.e. for every $N \geq 0$, there exists a constant C_N such that

$$|\widehat{\varphi f}(\boldsymbol{\xi})| \leq C_N \left(1 + ||\boldsymbol{\xi}||\right)^{-N}, \tag{3.31}$$

for all possible choices of $\boldsymbol{\xi} \to \infty$.

If $\mathbf{x}_0 \in$ sing supp(f), condition (3.31) may fail in some directions of $\boldsymbol{\xi}$ and hold in some others. This observation is key to the idea of "microlocal" description of singularities of f, characterized by the notion of the wavefront set of f.

Definition 3.10. The distribution $f \in \mathcal{D}'(\Omega)$ is called *microlocally smooth* at $(\mathbf{x}_0, \boldsymbol{\xi}_0) \in \Omega \times \mathbb{R}^2 \backslash \{\mathbf{0}\}$ if there exists $\varphi \in \mathcal{D}(\Omega)$ with $\varphi(\mathbf{x}_0) \neq 0$ and an open cone Γ containing $\boldsymbol{\xi}_0$ such that relation (3.31) holds for all $N \geq 0$ and every $\boldsymbol{\xi} \in \Gamma$. The *wavefront set of f*, denoted by $\mathrm{WF}(f)$, is the complement of the set of all $(\mathbf{x}, \boldsymbol{\xi})$ at which f is microlocally smooth.

3.4.2 *The main results*

In this section, we describe the results obtained by Sherson (2015). To the best of our knowledge, it is the only work where microlocal properties of the VLT have been investigated in formally determined setups (i.e. when the data set has two degrees of freedom) with V-line vertices inside the image domain. We are not aware of any studies about the microlocal properties of the corresponding CRTs in dimensions $d \geq 3$.

Sherson (2015) reported multiple interesting findings, including a study of injectivity and inversion formulas of VLTs in several geometric setups, investigation of their microlocal properties, and thorough numerical simulations. Here, we discuss the portion of that work dedicated to microlocal analysis of the VLT in slab geometry, where the transform $\mathcal{B}$ is defined and analyzed in the distributional sense. In particular, it includes a description of the propagation of singularities from f to $\mathcal{B}f$ and (using the inversion of $\mathcal{B}$) from $\mathcal{B}f$ to f.

The main building block of all statements discussed henceforth is the following characterization of the propagation of singularities by the divergent beam transform $\mathcal{X}_{\boldsymbol{\gamma}}$ defined by

$$\mathcal{X}_{\boldsymbol{\gamma}} f(\mathbf{x}) \doteq \int_0^\infty f(\mathbf{x} + t\boldsymbol{\gamma})\, dt.$$

Theorem 3.14 (Sherson, 2015). *Let $f \in \mathcal{D}'(\mathbb{R}^2)$. Then,*

$$\mathrm{WF}(\mathcal{X}_{\boldsymbol{\gamma}} f) \subseteq \mathrm{WF}(f) \cup \{(\boldsymbol{x} - t\boldsymbol{\gamma}, \boldsymbol{\xi}) \mid (\boldsymbol{x}, \boldsymbol{\xi}) \in \mathrm{WF}(f),\, \boldsymbol{\xi} \in \boldsymbol{\gamma}^\perp,\, t > 0\}. \quad (3.32)$$

In other words, in addition to the true singularities $\mathrm{WF}(f)$ of function f, its divergent beam transform data may also include a set of ancillary singularities, which start at the points where f has singularities in the direction normal to $\boldsymbol{\gamma}$ and propagate in the direction $-\boldsymbol{\gamma}$.

Combining the above statement with the definition of the transform $\mathcal{B}$ and its inverse $\mathcal{B}^{-1}$ leads to the following result.

Theorem 3.15 (Sherson, 2015). *Let $f \in \mathcal{E}'(\mathbb{R}^2)$. Then,*

$$\mathrm{WF}(\mathcal{B}f) \subseteq \mathrm{WF}(f) \cup \{(\boldsymbol{x} - t\boldsymbol{u}, \boldsymbol{\xi}) \mid (\boldsymbol{x}, \boldsymbol{\xi}) \in \mathrm{WF}(f),\, \boldsymbol{\xi} \in \boldsymbol{u}^\perp,\, t > 0\}$$
$$\cup \{(\boldsymbol{x} - t\boldsymbol{v}, \boldsymbol{\xi}) \mid (\boldsymbol{x}, \boldsymbol{\xi}) \in \mathrm{WF}(f),\, \boldsymbol{\xi} \in \boldsymbol{v}^\perp,\, t > 0\} \quad (3.33)$$

and

$$\mathrm{WF}(f) \subseteq \mathrm{WF}(\mathcal{B}f)$$
$$\cup \left\{ \left(\boldsymbol{x} - t(\boldsymbol{u} + \boldsymbol{v}), \boldsymbol{\xi} \right) \mid (\boldsymbol{x}, \boldsymbol{\xi}) \in \mathrm{WF}(\mathcal{B}f), \, \boldsymbol{\xi} \in (\boldsymbol{u} + \boldsymbol{v})^{\perp}, \, t > 0 \right\}, \quad (3.34)$$

$$\mathrm{WF}(f) \subseteq \mathrm{WF}(\mathcal{B}f)$$
$$\cup \left\{ \left(\boldsymbol{x} + t(\boldsymbol{u} + \boldsymbol{v}), \boldsymbol{\xi} \right) \mid (\boldsymbol{x}, \boldsymbol{\xi}) \in \mathrm{WF}(\mathcal{B}f), \, \boldsymbol{\xi} \in (\boldsymbol{u} + \boldsymbol{v})^{\perp}, \, t > 0 \right\}. \quad (3.35)$$

Moreover, if $(\boldsymbol{x}_0, \boldsymbol{\xi}_0) \in \mathrm{WF}(f) \setminus \mathrm{WF}(\mathcal{B}f)$, *then* $\boldsymbol{\xi}_0 \in (\boldsymbol{u} + \boldsymbol{v})^{\perp}$, *and* $\boldsymbol{x}_0$ *must lie on some line segment* $\boldsymbol{x}_0 + I\boldsymbol{w}$, *for which* $(\boldsymbol{x}_0 + I\boldsymbol{w}) \times \{\boldsymbol{\xi}_0\} \subseteq \mathrm{WF}(f)$ *and whose endpoints lie in* $\mathrm{WF}(\mathcal{B}f)$.

Let us discuss the above statements in relation to the artifacts that appear in the numerical reconstructions presented in this chapter.

Formula (3.33) indicates that in addition to the true singularities of the image, i.e. $\mathrm{WF}(f)$, the VLT data may also include two sets of ancillary singularities. They start at the points where the image f has singularities in the direction normal to $\mathbf{u}$ or $\mathbf{v}$ and propagate in the direction $-\mathbf{u}$ or $-\mathbf{v}$, respectively. Those added singularities can be seen, for example, in Figure 3.2(f) and (after an inversion formula is applied to the data) in Figure 3.2(e).

Formula (3.34) indicates that in addition to the true singularities $\mathrm{WF}(\mathcal{B}f)$ contained in the VLT data set, the wavefront set $\mathrm{WF}(f)$ of the reconstructed image f may also include an ancillary set of singularities. They start at the points where the data $\mathcal{B}f$ has singularities in the direction normal to $\mathbf{u} + \mathbf{v}$ and propagate in the direction $-(\mathbf{u} + \mathbf{v})$. These extra singularities become more pronounced in cases where the noise is added to the VLT data, e.g. see Figure 3.9.

Remark 3.12. All inversion formulas for $\mathcal{B}$ presented in this chapter involve an integration (i.e. a divergent beam transform) of the data in the direction $\mathbf{u} + \mathbf{v}$. At the same time, one can easily get analogous formulas integrating in the direction $-(\mathbf{u} + \mathbf{v})$. Formula (3.35) is the "symmetric version" of (3.34) and reflects the aforementioned fact.

Remark 3.13. Note that formulas (3.33)–(3.35) are containment relations and not equalities. Therefore, it is theoretically possible that certain true singularities of the object are not included in its transform data.

The last part of Theorem 3.15 states that the VLT data with a fixed axis of symmetry and a fixed opening angle allows the recovery of singularities of f at all points and in all directions, except (possibly) in the direction normal to the axis of symmetry in some very restrictive cases. An explicit example by Sherson (2015) shows that $\mathrm{WF}(f) \setminus \mathrm{WF}(\mathcal{B}f)$ can indeed be nonempty.

3.5   Additional Remarks

- Another inversion formula for the VLT in slab geometry is discussed in Chapter 5 as a special case of the inversion of the star transform (see Corollary 5.2).

- Inversion formulas for the unweighted VLT $\mathcal{B}$ and the signed VLT $\mathcal{B}^-$, equivalent to those presented in this chapter, have been obtained by Katsevich and Krylov (2013) using the techniques of first-order partial differential equations. For more on this topic, see Section 4.2.1.

- A generalization of the CRT in $\mathbb{R}^d$ discussed in Section 3.2.2 was considered by Palamodov (2017). The family of cones used in that transform is the same (arbitrary vertex locations, fixed opening angle, and fixed axis of symmetry), but integrals are computed with a weight $\omega(\mathbf{x}) = |\mathbf{x}|^{-k}$, $k < d - 1$, i.e. a power of the distance from the variable location on the cone to the vertex of the cone. This weighted CRT is expressed as a convolution with a distribution supported on the cone, and the reconstruction is given in terms of the cone transform with another weight. The technique uses a connection of the cone transform with the fundamental solution to the wave operator.

- In a recent work by Baines (2021), the connection between the wave operator and the cone transform was used to devise a new range description for the unweighted VLT and CRT acting on $C_0^\infty(\mathbb{R}^d)$ in the slab geometry. The solution utilizes the fact that the application of a modified wave operator to the VLT data $g(x,t)$ produces (up to a constant multiple) the derivative $f_t'(\mathbf{x},t)$ of the image function f with respect to the "vertical" variable t (measured along the axis of symmetry of the cones). Therefore, that quantity has compact support, and its integral over $\mathbb{R}$ with respect to t is equal to zero. It is also easy to see that the support of the VLT data is contained in a half plane. It is shown in the paper that the above three conditions are also sufficient for g to be in the range of the VLT. The result is generalized to CRTs in arbitrary even dimensions $2k$ by substituting the modified wave operator in the above description by its power k. A slightly altered version of the result also holds for the CRTs in odd dimensions. Here, twice the higher powers of the modified wave operator are applied to a weighted conical transform of the CRT data to generate $f'_t(\mathbf{x}, t)$. The ensuing range conditions are analogous to those in even dimensions.

- Various practical aspects of inverting the VLT in slab geometry were discussed in a thorough paper by Walker and O'Sullivan (2019). Motivated by the applications related to coherent scatter X-ray imaging (recall Section 2.3.1), the authors demonstrate the need for the study of the 2D restrictions of the VLT using only two directions. They present a comprehensive analysis of different known inversion formulas, identify issues with their numerical implementations, and compare their performance. The authors also leverage the 2D Fourier transform to derive a new, computationally efficient inversion algorithm for the VLT.

- In a follow-up work, Walker and O'Sullivan (2021) presented a generalized iterative algorithm for the inversion of the VLT in 2D. The algorithm incorporates multiple source locations, scatter angles, and transmission measurements and allows the joint reconstruction of the scatter density or total attenuation. The efficiency of the proposed technique was demonstrated using numerical examples, including image reconstructions from noisy single-scatter measurements with missing data.

- An interesting concept of nonreciprocal VLTs (NVLT) was considered bt Florescu *et al.* (2018). As opposed to the standard VLTs, the values of which do not change when one flips the direction of travel along a V-line, the NVLTs produce different outcomes when the direction of travel is reversed (i.e. the locations of the photon emitter and receiver are interchanged). Some physical motivations for such transforms include SSXT, where the photon changes its energy after scattering, and optical tomography with fluorescent contrast agents, where the photon is absorbed by a fluorophore molecule and then re-emitted in a different direction and at a different frequency. It is shown in the paper that by utilizing the nonreciprocity of the data function, one can simultaneously reconstruct the attenuation coefficients of the medium both at the excitation and fluorescence frequencies, as well as the concentration of the contrast agent. The solution is obtained by inverting a three-ray (nonreciprocal) star transform using mainly algebraic techniques.

- An attenuated version (with an exponential weight) of the CRT in $\mathbb{R}^d$ discussed in Section 3.2.2 was considered by Gouia-Zarrad and Moon (2018). The authors derived an inversion formula for that transform and tested it in 2D numerical simulations. The singular value decomposition of that transform was derived later by Jeon and Moon (2020).

Chapter 4

V-line Transforms in Curvilinear Geometry

4.1 Rotation-Invariant Setups

The VLTs considered in the previous chapter mapped a function to its integrals over *translation-invariant* families of V-lines. As a result, several inversion formulas were obtained by utilizing Fourier transform techniques. In this section, we consider transforms using *rotation-invariant* families of V-lines. The conventional wisdom in integral geometry suggests that these transforms should be invertible using Fourier series expansions (e.g. see Cormack, 1963, 1964; Ambartsoumian *et al.*, 2010; Ambartsoumian and Moon, 2013; Ambartsoumian and Krishnan, 2015).

We start with a circular setup shown in Figure 2.5, which was motivated by the experimental ideas of Florescu *et al.* (2010, 2009) applied to a rectangular slab. Namely, we assume that the image domain is supported inside the unit disc. The V-lines, corresponding to paths of single scattered photons, enter the disc along the normals to its boundary. The receivers are collimated in a way that all V-lines have the same opening angle 2β. The resulting family of V-lines is clearly rotation-invariant and two-dimensional.

Let us parametrize each V-line by the signed distance $t \in [-1, 1]$ of its vertex to the origin and the polar angle $\theta \in [0, 2\pi]$ of the vertex (or, equivalently, of the V-line's point of entry into the disc) (see Figure 4.1). For example, in Figure 4.1, the V-line A_1BC_1 has parameters (θ, t), while the V-line A_2BC_2 has parameters $(\theta + \pi, -t)$. We denote the V-line with parameters (θ, t) by $L(\theta, t)$ and the corresponding value of VLT of f by $\mathcal{B}f(\theta, t)$.

If the VLT data is known for all $t \in [-1, 1]$ and $\theta \in [0, 2\pi]$, then for a certain restricted class of functions f, one can invert the VLT by recovering

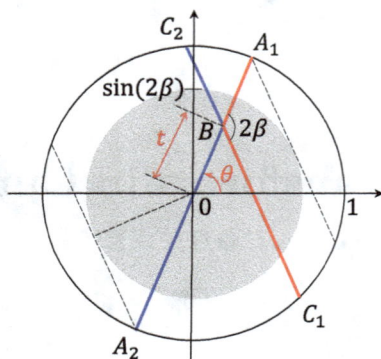

Fig. 4.1   Parametrization of V-lines in the unit disc: The two parallel lines tangent to the small circle are the lines farthest from the origin in the Radon data set.

the classical Radon transform (RT) of the image function and then applying a standard approach to invert the RT (Ambartsoumian, 2012). Namely, a simple geometric observation shows that

$$\mathcal{B}f(\theta, t) + \mathcal{B}f(\theta + \pi, -t) = \mathcal{R}f(\theta - \pi/2, 0) + \mathcal{R}f(\theta + \pi/2 - 2\beta, t \sin(2\beta)),$$

where $\mathcal{R}f$ denotes the RT of f in $\mathbb{R}^2$ (see formula (2.31)):

$$\mathcal{R}f(\theta, t) = \int_{\mathbf{x}\cdot\hat{\mathbf{n}}=t} f \, dl$$

and $\hat{\mathbf{n}} = (\cos\theta, \sin\theta)$.

Note that $\mathcal{R}f(\theta - \pi/2, 0) = \mathcal{B}f(\theta, -1)$. Then, one can recover the Radon data from the VLT data using the following formula:

$$\mathcal{R}f(\theta, t \sin(2\beta)) = \mathcal{B}f(\theta + 2\beta - \pi/2, t) + \mathcal{B}f(\theta + 2\beta + \pi/2, -t)$$
$$- \mathcal{B}f(\theta + 2\beta - \pi/2, -1). \tag{4.1}$$

Remark 4.1. Note that the approach described above is applicable with a trivial modification to the case of the weighted VLT $\mathcal{B}^c f$, integrating with a constant weight c_1 along one ray (i.e. before scattering) and with another constant weight c_2 along the other ray (i.e. after scattering). Such transforms appear in mathematical models of SSXT employing high-energy X-rays (recall Section 2.3.2).

Remark 4.2. Using the full range of values of both θ and t, one can obtain the Radon data only for lines that pass through the disc of radius $\sin(2\beta)$

(see the shaded region in Figure 4.1). The Radon transform with such (interior) data is not injective (Natterer, 2001b), unless the support of f is restricted to that smaller disc. One may argue that if f is zero outside of the smaller disc, then (in applications) there will be no scattering and hence no VLT data with vertices there. That can be remedied by placing the object in a medium of appropriate thickness with known properties before making the measurements and then subtracting from the measured data the portion of VLT corresponding to the layer of known material.

The data set used in the above setup includes two different V-lines with a vertex at each point of the disc. It is natural to ask if one can invert the VLT from half of that data, i.e. using $\mathcal{B}f(\theta, t)$ for $\theta \in [0, 2\pi]$ and $t \in [0, 1]$. A positive answer to that question was given by Ambartsoumian and Moon (2013) using Fourier series techniques.

For brevity, let us denote $g(\theta, t) \doteq \mathcal{B}f(\theta, t)$ and let $f(\phi, r)$ be the image function in polar coordinates. Then, the Fourier series of $f(\phi, r)$ and $g(\theta, t)$ with respect to their angular variables can be written as follows:

$$f(\phi, r) = \sum_{n=-\infty}^{\infty} f_n(r)\, e^{in\phi}, \qquad g(\theta, t) = \sum_{n=-\infty}^{\infty} g_n(t)\, e^{in\theta},$$

where the Fourier coefficients are given by

$$f_n(r) = \frac{1}{2\pi} \int_0^{2\pi} f(\phi, r)\, e^{-in\phi}\, d\phi, \qquad g_n(t) = \frac{1}{2\pi} \int_0^{2\pi} g(\theta, t)\, e^{-in\theta}\, d\theta.$$

Due to the rotation invariance of the family of V-lines used in our data, the Fourier series expansion diagonalizes our operator. In other words, the nth Fourier coefficient of $\mathcal{B}f$ depends only on the nth Fourier coefficient of f. The exact formula of this relation was established by Ambartsoumian and Moon (2013), who then used it to invert $\mathcal{B}$ by expressing f_n in terms of g_n.

Theorem 4.1. *Let f be supported inside the disc $D(0, 1)$. Then, we have*

$$\mathcal{M}f_n(s) = \frac{\mathcal{M}g_n(s-1)}{1/(s-1) + \mathcal{M}h_n(s-1)}, \qquad \Re(s) > 1, \tag{4.2}$$

where $\mathcal{M}F$ denotes the Mellin transform of function F:

$$\mathcal{M}F(s) = \int_0^{\infty} p^{s-1} F(p)\, dp$$

and $h_n(t)$ is an explicitly defined (albeit cumbersome), elementary function of t (see Ambartsoumian and Moon, 2013, for the exact expression).

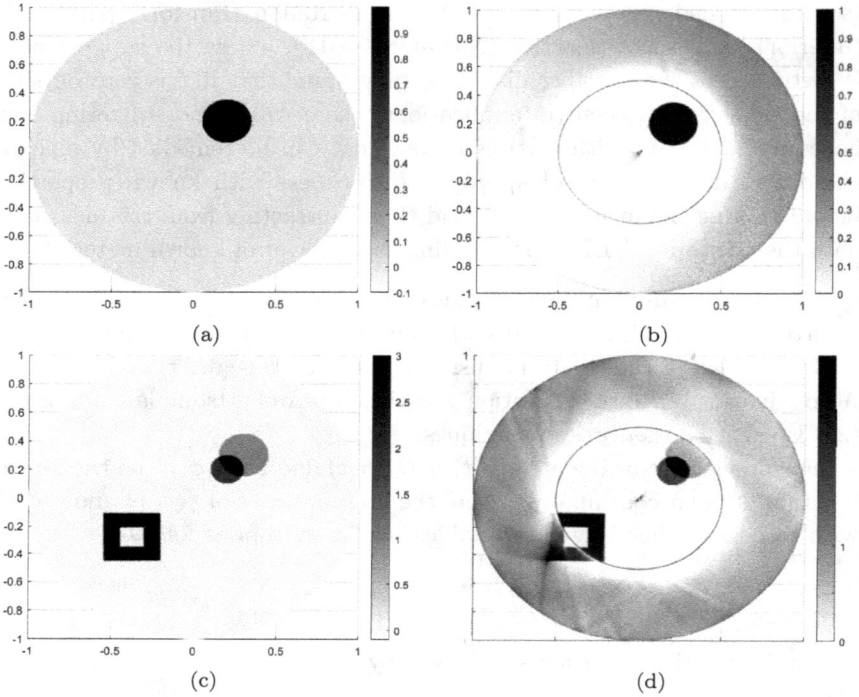

Fig. 4.2   Numerical simulations of VLT and its inversion in circular geometry of data acquisition. Both the phantom and the VLT data are discretized on a 400×400 grid. The inversion employs the algorithms developed by Ambartsoumian and Roy (2016). (a) and (c) are the phantoms, and (b) and (d) are their reconstructions. The reconstructed images have streak artifacts at the origin and along the circle of radius $\sin(2\beta) = 0.5$. These are due to the abrupt cut in the (incomplete) VLT data at $t = 0$ and $t = 1$, respectively. See Ambartsoumian and Roy (2016) for more details. The blurring of the image features outside of the small circle is akin to the standard artifact of limited angle reconstructions (e.g. see Krishnan and Quinto, 2015). *Implementation of these numerical simulations is courtesy of Souvik Roy.*

By applying the standard inverse of the Mellin transform, one can recover the f_ns from the g_ns, which finishes the process of inverting the VLT in a disc from (radially) half of the data.

A numerical implementation of the inversion procedure based on (4.2) was presented by Ambartsoumian and Roy (2016). Two examples using the algorithm developed in that paper are presented in Figure 4.2. In both simulations, the opening angle of the V-lines is $2\beta = 5\pi/6$, and the VLT data $g(\theta, t) = \mathcal{B}f(\theta, t)$ is known for $\theta \in [0, 2\pi]$ and $t \in [0, 1]$.

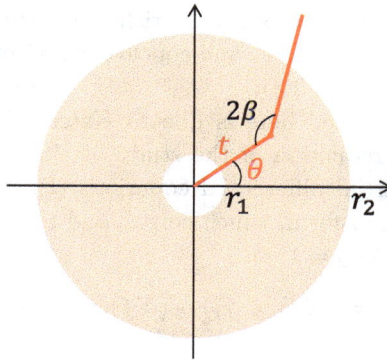

Fig. 4.3   Parametrization of polar broken rays in the unit disc.

Another interesting VLT, using a rotationally invariant family of V-lines, is considered by Sherson (2015). This operator, called *polar broken ray transform* (**PBRT**), maps the functions supported inside a closed annulus $A(r_1, r_2) = \overline{D(0, r_2)} \setminus D(0, r_1)$, $0 < r_1 < r_2$, to their integrals along the following trajectories. Each broken ray $L(\theta, t)$ starts at the origin and travels a distance $t > 0$ along the radius vector with a polar angle $\theta \in [0, 2\pi]$ before breaking at a fixed angle $2\beta \in (\pi/2, \pi)$ (see Figure 4.3).

A Fourier series approach similar to the one described above is used by Sherson (2015) to derive an inversion of PBRT. It is shown that the Fourier coefficients f_n and g_n are connected through a Volterra integral equation of the second kind, for which a series solution is derived. The efficiency of the method is demonstrated through various numerical simulations. The author also studied the microlocal properties of PBRT, describing in detail how that transform propagates singularities of the image function.

4.2   Configurations with Focal Points

4.2.1   *Setups with two directions*

In this section, we discuss the VLTs in setups where their branch directions $\gamma_1(\mathbf{x})$ and $\gamma_2(\mathbf{x})$ correspond to vector fields whose integral curves are straight lines. For the V-line branches representing the incident field of radiation, the above condition is necessary (recall Remark 2.3) and cannot be considered a restriction. The requirement for the V-line branches corresponding to the scattered beam is satisfied in setups with linear arrays[1] of detectors, as well as for convex and concave arrays of detectors discussed in Section 2.3.1 (see Figure 2.8).

The approach discussed here is due to Katsevich and Krylov (2013) and is based on the reduction of the study of VLTs to the treatment of certain first-order partial differential equations (PDEs). We consider two special cases of the transform, which correspond to the weight parameters $c_1 = c_2 = 1$ and $c_1 = -c_2 = 1$, i.e.

$$\mathcal{B}f(\mathbf{x}, \Gamma) \doteq (\mathcal{X}f)(\mathbf{x}, \boldsymbol{\gamma}_1) + (\mathcal{X}f)(\mathbf{x}, \boldsymbol{\gamma}_2), \quad \mathbf{x} \in \mathbb{R}^2,$$

and

$$\mathcal{B}^- f(\mathbf{x}, \Gamma) \doteq (\mathcal{X}f)(\mathbf{x}, \boldsymbol{\gamma}_1) + (\mathcal{X}f)(\mathbf{x}, \boldsymbol{\gamma}_2), \quad \mathbf{x} \in \mathbb{R}^2,$$

where $\Gamma = [\boldsymbol{\gamma}_1, \boldsymbol{\gamma}_2]$ is a 2×2 matrix function with columns $\boldsymbol{\gamma}_1(\mathbf{x}), \boldsymbol{\gamma}_2(\mathbf{x}) \in \mathbb{S}^1$.

Let us start by introducing some notations. For indices $i, j \in \{1, 2\}$, denote

$$c_{ij}(\mathbf{x}) \doteq \boldsymbol{\gamma}_i(\mathbf{x}) \cdot \boldsymbol{\gamma}_j(\mathbf{x}) \quad \text{and} \quad c_{ij}^{\perp}(\mathbf{x}) \doteq \boldsymbol{\gamma}_i(\mathbf{x}) \cdot \boldsymbol{\gamma}_j^{\perp}(\mathbf{x}), \qquad (4.3)$$

where $\boldsymbol{\gamma}_j^{\perp}(\mathbf{x})$ is the unit vector obtained from $\boldsymbol{\gamma}_j(\mathbf{x})$ by a 90° counterclockwise rotation. Writing in vector components, we have $\boldsymbol{\gamma}_j(x) = (a_j(\mathbf{x}), b_j(\mathbf{x}))$ and $\boldsymbol{\gamma}_j^{\perp}(\mathbf{x}) = (-b_j(x), a_j(x))$. To simplify the formulas, we often drop the notation of dependence on $\mathbf{x}$.

Let $D_{\boldsymbol{\gamma}_j} \doteq \boldsymbol{\gamma}_j \cdot \nabla$ and $D_{\boldsymbol{\gamma}_j}^{\perp} \doteq \boldsymbol{\gamma}_j^{\perp} \cdot \nabla$ denote the directional derivatives in the direction of vectors $\boldsymbol{\gamma}_j$ and $\boldsymbol{\gamma}_j^{\perp}$, respectively. When applied to a vector field, we treat these operators as directional derivatives applied separately to each of its components.

Next, we deduce several auxiliary identities, which will be helpful in the derivation of the main results.

One can verify by direct calculation that for $i, j \in \{1, 2\}$ and $i \neq j$,

$$D_{\boldsymbol{\gamma}_i} = c_{ij} D_{\boldsymbol{\gamma}_j} + c_{ij}^{\perp} D_{\boldsymbol{\gamma}_j}^{\perp}. \qquad (4.4)$$

Also, since in our setups all integral curves of the vector fields $\boldsymbol{\gamma}_1$ and $\boldsymbol{\gamma}_2$ are straight lines,

$$D_{\boldsymbol{\gamma}_i} \boldsymbol{\gamma}_i(\mathbf{x}) \equiv 0 \quad \text{and} \quad D_{\boldsymbol{\gamma}_i} \boldsymbol{\gamma}_i^{\perp}(\mathbf{x}) \equiv 0, \quad i = 1, 2. \qquad (4.5)$$

Note that $\|D_{\boldsymbol{\gamma}_i} \boldsymbol{\gamma}_i(\mathbf{x})\|$ represents the curvature of the corresponding integral curve at point $\mathbf{x}$.

---

[1]Detectors on a linear (flat) array can be considered as focused to a point at infinity.

Several interesting observations follow from the fact that $\boldsymbol{\gamma}_i$s are unit vector fields. Namely, since $a_i^2 + b_i^2 = 1$, we have

$$a_i \frac{\partial a_i}{\partial x_j} + b_i \frac{\partial b_i}{\partial x_j} = 0, \quad \forall i,j \in \{1,2\}. \tag{4.6}$$

Using the above relation, it is easy to verify that

$$D_{\boldsymbol{\gamma}_i^\perp} \boldsymbol{\gamma}_i = \operatorname{div}(\boldsymbol{\gamma}_i)\, \boldsymbol{\gamma}_i^\perp. \tag{4.7}$$

For brevity, we denote $l_i(\mathbf{x}) \doteq -\operatorname{div}(\boldsymbol{\gamma}_i)(\mathbf{x})$.

In the case of a *flat* detector array, the corresponding vector field $\boldsymbol{\gamma}_i$ is constant; therefore, $l_i(\mathbf{x}) \equiv 0$. If the detector is *concave*[2] with a focus at the point $\mathbf{x}_0$ (see Figure 2.8), then

$$\boldsymbol{\gamma}_i(\mathbf{x}) = \frac{\mathbf{x} - \mathbf{x}_0}{\|\mathbf{x} - \mathbf{x}_0\|}$$

and $l_i(\mathbf{x}) = -\|\mathbf{x} - \mathbf{x}_0\|^{-1}$. If the detector is *convex*[3] with a focus at the point $\mathbf{x}_0$, then

$$\boldsymbol{\gamma}_i(\mathbf{x}) = \frac{\mathbf{x}_0 - \mathbf{x}}{\|\mathbf{x} - \mathbf{x}_0\|}$$

and $l_i(\mathbf{x}) = \|\mathbf{x} - \mathbf{x}_0\|^{-1}$.

One can also verify by direct calculation the following identity:

$$D_{\boldsymbol{\gamma}_j} D_{\boldsymbol{\gamma}_j}^\perp - D_{\boldsymbol{\gamma}_j}^\perp D_{\boldsymbol{\gamma}_j} = l_j D_{\boldsymbol{\gamma}_j}^\perp. \tag{4.8}$$

Let us apply the directional derivative $D_{\boldsymbol{\gamma}_i}$ to the signed VLT $\mathcal{B}^- f$. Using identities (4.4) and (4.5), we obtain

$$D_{\boldsymbol{\gamma}_i}(\mathcal{B}^- f)(\mathbf{x}) = -f(\mathbf{x}) + c_{ij}(\mathbf{x})\, f(\mathbf{x}) - c_{ij}^\perp(\mathbf{x})\, D_{\boldsymbol{\gamma}_j}^\perp(\mathcal{X} f)(\mathbf{x}, \boldsymbol{\gamma}_j) \tag{4.9}$$

$$= [-1 + c_{ij}(\mathbf{x})]\, f(\mathbf{x}) - c_{ij}^\perp(\mathbf{x})\, J_j(\mathbf{x}),$$

where

$$J_j(\mathbf{x}) \doteq D_{\boldsymbol{\gamma}_j}^\perp(\mathcal{X} f)(\mathbf{x}, \boldsymbol{\gamma}_j). \tag{4.10}$$

Also, utilizing the relations (4.4), (4.5), and (4.7), we get

$$D_{\boldsymbol{\gamma}_j} c_{ij} = [D_{\boldsymbol{\gamma}_j} \boldsymbol{\gamma}_i] \cdot \boldsymbol{\gamma}_j = [(c_{ji} D_{\boldsymbol{\gamma}_i} + c_{ji}^\perp D_{\boldsymbol{\gamma}_i}^\perp)\, \boldsymbol{\gamma}_i] \cdot \boldsymbol{\gamma}_j$$

$$= -c_{ji}^\perp l_i c_{ji}^\perp = -l_i \left(c_{ji}^\perp\right)^2 \tag{4.11}$$

---

[2]In this case, $\mathbf{x}_0$ acts like a "source"; therefore, $\operatorname{div}(\boldsymbol{\gamma}_i) > 0$.
[3]In this case, $\mathbf{x}_0$ acts like a "sink"; therefore, $\operatorname{div}(\boldsymbol{\gamma}_i) < 0$.

and

$$D_{\boldsymbol{\gamma}_j} c_{ij}^{\perp} = [D_{\boldsymbol{\gamma}_j} \boldsymbol{\gamma}_i] \cdot \boldsymbol{\gamma}_j^{\perp} = -c_{ji}^{\perp} l_i \, \boldsymbol{\gamma}_i^{\perp} \boldsymbol{\gamma}_j^{\perp} = -l_i \, c_{ji}^{\perp} c_{ij}. \tag{4.12}$$

Applying the directional derivative $D_{\boldsymbol{\gamma}_j}$ to both sides of equation (4.9) and taking advantage of the previous two relations, we get

$$D_{\boldsymbol{\gamma}_j} D_{\boldsymbol{\gamma}_i} (\mathcal{B}^- f)(\mathbf{x}) = -l_i(\mathbf{x}) \left[c_{ji}^{\perp}(\mathbf{x}) \right]^2 f(\mathbf{x}) + [-1 + c_{ij}(\mathbf{x})] D_{\boldsymbol{\gamma}_j} f(\mathbf{x})$$
$$+ l_i(\mathbf{x}) \, c_{ji}^{\perp}(\mathbf{x}) \, c_{ij}(\mathbf{x}) J_j(\mathbf{x}) - c_{ij}^{\perp}(\mathbf{x}) \, D_{\boldsymbol{\gamma}_j} J_j(\mathbf{x}). \tag{4.13}$$

The left-hand side of equation (4.13) can be computed from the VLT data. Our goal is to eliminate integrals from the right-hand side of (4.13) and obtain a PDE with respect to the unknown function f.

Using relations (4.8) and (4.10), we express

$$D_{\boldsymbol{\gamma}_j} J_j(\mathbf{x}) = \left(D_{\boldsymbol{\gamma}_j}^{\perp} D_{\boldsymbol{\gamma}_j} + l_j(\mathbf{x}) D_{\boldsymbol{\gamma}_j}^{\perp} \right) (\mathcal{X} f)(\mathbf{x}, \boldsymbol{\gamma}_j)$$
$$= -D_{\boldsymbol{\gamma}_j}^{\perp} f(\mathbf{x}) + l_j(\mathbf{x}) J_j(\mathbf{x}). \tag{4.14}$$

Combining equations (4.10), (4.13), and (4.14), we arrive at the following relation:

$$D_{\boldsymbol{\gamma}_j} D_{\boldsymbol{\gamma}_i} (\mathcal{B}^- f)(\mathbf{x}) = -l_i(\mathbf{x}) \left[c_{ji}^{\perp}(\mathbf{x}) \right]^2 f(\mathbf{x}) + [-1 + c_{ij}(\mathbf{x})] D_{\boldsymbol{\gamma}_j} f(\mathbf{x})$$
$$+ c_{ij}^{\perp}(\mathbf{x}) \, D_{\boldsymbol{\gamma}_j}^{\perp} f(\mathbf{x}) + \left[l_i(\mathbf{x}) \, c_{ji}^{\perp}(\mathbf{x}) \, c_{ij}(\mathbf{x}) \right.$$
$$\left. - c_{ij}^{\perp}(\mathbf{x}) \, l_j(\mathbf{x}) \right] [(-1 + c_{ij}(\mathbf{x})) f(\mathbf{x})$$
$$- D_{\boldsymbol{\gamma}_i} (\mathcal{B}^- f)(\mathbf{x})] / c_{ij}^{\perp}(\mathbf{x}).$$

After some simplification, we obtain the following first-order PDE:

$$[l_j(\mathbf{x}) - l_i(\mathbf{x})] [1 - c_{ij}(\mathbf{x})] f(\mathbf{x}) + (D_{\boldsymbol{\gamma}_i} - D_{\boldsymbol{\gamma}_j}) f(\mathbf{x})$$
$$= D_{\boldsymbol{\gamma}_j} D_{\boldsymbol{\gamma}_i} (\mathcal{B}^- f)(\mathbf{x}) - [l_i(\mathbf{x}) c_{ij}(\mathbf{x}) + l_j(\mathbf{x})] D_{\boldsymbol{\gamma}_i} (\mathcal{B}^- f)(\mathbf{x}). \tag{4.15}$$

In particular, in the case of flat arrays of detectors (i.e. when $\boldsymbol{\gamma}_1$ and $\boldsymbol{\gamma}_2$ are constant vector fields), equation (4.15) simplifies into the following:

$$(D_{\boldsymbol{\gamma}_1} - D_{\boldsymbol{\gamma}_2}) f(\mathbf{x}) = D_{\boldsymbol{\gamma}_2} D_{\boldsymbol{\gamma}_1} (\mathcal{B}^- f)(\mathbf{x}). \tag{4.16}$$

Both equations can be solved by the method of characteristics (e.g. see Chechkin and Goritsky, 2009). If the detector arrays have focal points located at $\mathbf{x}_0^{(1)}$ and $\mathbf{x}_0^{(2)}$, then the characteristic curves of (4.15) correspond to ellipses with foci at $\mathbf{x}_0^{(1)}$ and $\mathbf{x}_0^{(2)}$. If one of the detector arrays is flat and collimated along $\boldsymbol{\gamma}_1$ while the other one has a focal point at $\mathbf{x}_0^{(2)}$, then the characteristic curves of (4.15) correspond to parabolas with a focus at $\mathbf{x}_0^{(2)}$ and an axis of symmetry along $\boldsymbol{\gamma}_1$. The above statements follow from the reflection properties of ellipses and parabolas, respectively. The characteristic curves of (4.16) are straight lines parallel to $\boldsymbol{\gamma}_1 - \boldsymbol{\gamma}_2$.

If it is known that f is supported inside a convex domain U and $(\mathcal{B}^- f)(\mathbf{x})$ is known for every $\mathbf{x} \in U$, then one can use "$f = 0$ on ∂U" as a boundary condition for solving the above PDEs.

Remark 4.3. In slab geometry (i.e. when $\boldsymbol{\gamma}_1$ and $\boldsymbol{\gamma}_2$ are constant vector fields), the PDE approach outlined above works also for the inversion of the transform $\mathcal{B}$ corresponding to uniform weights $c_1 = c_2 = 1$. The resulting inversion formulas for both transforms coincide with the formulas obtained in the previous chapter using cone differentiation and integration.

Remark 4.4. Explicit, closed-form solutions of equation (4.15) were obtained by Sherson (2015) for various combinations of convex, concave, and flat detectors using appropriate changes to the curvilinear coordinate system. Giving distributional meaning to those solutions, the author also described the propagation of singularities along parabolas, hyperbolas, and ellipses by the ensuing inversion operators.

4.2.2 *Setups with three directions*

Let us now assume that the VLT data is measured using three sets of detectors, i.e. one has access to $g_{ij}(\mathbf{x}) \doteq (\mathcal{B}^- f)(\mathbf{x}, \Gamma_{ij})$, where $\Gamma_{ij} = [\boldsymbol{\gamma}_i, \boldsymbol{\gamma}_j]$ for all $1 \leq i, j \leq 3$, $i \neq j$.

Equation (4.9) implies that

$$D_{\boldsymbol{\gamma}_1} g_{12}(\mathbf{x}) = (-1 + c_{12}) f(\mathbf{x}) - c_{12}^{\perp} J_2(\mathbf{x}),$$
$$D_{\boldsymbol{\gamma}_3} g_{32}(\mathbf{x}) = (-1 + c_{32}) f(\mathbf{x}) - c_{32}^{\perp} J_2(\mathbf{x}).$$

Taking a linear combination of the above equations to eliminate the terms with $J_2(\mathbf{x})$ and utilizing the easily verifiable identity

$$c_{32}^{\perp} c_{12} + c_{21}^{\perp} c_{32} = -c_{13}^{\perp} \tag{4.17}$$

leads to an inversion formula for the VLT with three directions:

$$f(\mathbf{x}) = -\frac{c_{32}^{\perp} D_{\boldsymbol{\gamma}_1} g_{12}(\mathbf{x}) + c_{21}^{\perp} D_{\boldsymbol{\gamma}_3} g_{32}(\mathbf{x})}{c_{32}^{\perp} + c_{21}^{\perp} + c_{13}^{\perp}}. \tag{4.18}$$

Of course, similar relations can be obtained using other pairs of $g_{ij}(\mathbf{x})$.

Remark 4.5. Note that this inversion formula is local, i.e. to recover the value of the function at a point $\mathbf{x}$, one only needs to know its VLT data along the V-lines (in three directions) with vertices in the neighborhood of $\mathbf{x}$. This is another remarkable example showing that the GRTs with a "vertex" in the path of integration are qualitatively different from the classical GRTs integrating along smooth trajectories, where such a result is not possible in principle.

Building up on these results, Katsevich and Krylov (2013) also provided a range description of the VLT with three directions.

Theorem 4.2 (Katsevich and Krylov, 2013). *Given three functions* $g_i \in C^{\infty}(\mathbb{R}^2)$, $i = 1, 2, 3$, *define* $g_{i,j} \doteq g_i - g_j$, $i \neq j$. *Let* $U \subset \mathbb{R}^2$ *be an open convex set and all focal points* $\mathbf{x}_0^{(i)}$ *be outside the closure of* U. *Denote by* $R_i(\mathbf{x}) \doteq \{\mathbf{x} + t\boldsymbol{\gamma}_i, \forall t \geq 0\}$ *the ray emanating from* $\mathbf{x}$ *in the direction* $\boldsymbol{\gamma}_i$. *There exists* $f \in C_0^{\infty}(U)$ *such that*

$$g_{ij}(\mathbf{x}) = (\mathcal{B}^- f)(\mathbf{x}, \Gamma_{ij}), \quad 1 \leq i, j \leq 3, \ i \neq j, \tag{4.19}$$

if and only if

(1) *for all* $1 \leq i, j \leq 3$, $i \neq j$, *and* $\mathbf{x} \notin U$ *we have:*

    (a) $g_{ij}(\mathbf{x}) = 0$ *if the rays* $R_i(\mathbf{x})$ *and* $R_j(\mathbf{x})$ *do not intersect* U;

    (b) $D_{\boldsymbol{\gamma}_i} g_{ij}(\mathbf{x}) = 0$ *if* $R_i(\mathbf{x})$ *intersects* U, *but* $R_j(\mathbf{x})$ *does not intersect* U;

    (c) $c_{ik}^{\perp} D_{\boldsymbol{\gamma}_j} g_{jk}(\mathbf{x}) + c_{kj}^{\perp} D_{\boldsymbol{\gamma}_i} g_{ik}(\mathbf{x}) = 0$, *where* $k \neq i, j$, *and both* $R_i(\mathbf{x})$ *and* $R_j(\mathbf{x})$ *intersect* U;

(2) *the following relation holds for any* $\mathbf{x} \in \mathbb{R}^2$:

$$(D_{\boldsymbol{\gamma}_3} - D_{\boldsymbol{\gamma}_2}) D_{\boldsymbol{\gamma}_1} g_1 + (D_{\boldsymbol{\gamma}_1} - D_{\boldsymbol{\gamma}_3}) D_{\boldsymbol{\gamma}_2} g_2 + (D_{\boldsymbol{\gamma}_2} - D_{\boldsymbol{\gamma}_1}) D_{\boldsymbol{\gamma}_3} g_3$$

$$+ \frac{A_1 D_{\boldsymbol{\gamma}_1} g_1 + A_2 D_{\boldsymbol{\gamma}_2} g_2 + A_3 D_{\boldsymbol{\gamma}_3} g_3}{c_{32}^{\perp} + c_{21}^{\perp} + c_{13}^{\perp}} = 0, \tag{4.20}$$

where A_1, A_2, *and* A_3 *are explicitly defined (albeit cumbersome) functions of* $\mathbf{x}$ *(see Katsevich and Krylov, 2013, for the exact expressions).*

Remark 4.6. In the case of constant vector fields $\boldsymbol{\gamma}_i$, $i = 1, 2, 3$, the range condition (4.20) simplifies into the following:

$$(D_{\gamma_3} - D_{\gamma_2})D_{\gamma_1}g_1 + (D_{\gamma_1} - D_{\gamma_3})D_{\gamma_2}g_2 + (D_{\gamma_2} - D_{\gamma_1})D_{\gamma_3}g_3 = 0. \quad (4.21)$$

Remark 4.7. Theorem 4.2 assumes that the functions $g_i(\mathbf{x})$ (and hence their differences $g_{ij}(\mathbf{x})$) are known for all $\mathbf{x} \in \mathbb{R}^2$. If they are known only for $\mathbf{x} \in \operatorname{supp} f$, then (4.20) becomes a necessary range condition. This condition is local, so it can be verified for all $\mathbf{x}$, where $g_i(\mathbf{x})$ are known.

4.2.3 Additional remarks

- One can consider generalizations of the V-line transforms and the associated mathematical problems to Riemannian manifolds. To the best of our knowledge, it has been done only in the case of broken geodesics reflecting from a boundary (e.g. see Kurylev *et al.*, 2010).

- An interesting area of research in integral geometry is dedicated to the recovery of functions defined inside a compact domain from their integrals along piecewise-linear trajectories that reflect multiple times from the boundary of that domain. As it often happens in mathematics, this transform is also called a broken-ray transform, although it is quite different from the BRT mentioned in this book. For more details and interesting results in this field, we refer the reader to Hubenthal (2014, 2015); Ilmavirta (2013); Ilmavirta and Salo (2016).

Chapter 5

Star Transform

In this chapter, we study the star transform, which was introduced in the pioneering paper by Zhao *et al.* (2014) in relation to the mathematical models of SSOT and SSXT employing multiple sets of detectors (see Section 2.2.3). As opposed to the VLT, which has been extensively investigated by numerous authors, the star transform (with three or more branches) is a modestly studied subject, with only two publications on that topic (Zhao *et al.*, 2014; Ambartsoumian and Latifi, 2021).

Zhao *et al.* (2014) used Fourier analysis techniques to reduce the problem of inverting the star transform to solving an infinite system of linear equations. The latter is handled using various truncations of the system and fast iterative computations of the Tikhonov-regularized pseudo-inverses of the resulting finite-size rectangular matrices. The paper also discusses the stability of the proposed inversion algorithms, formulating an algebraic condition that is necessary and sufficient for the stable reconstruction of the image function. Several special geometric setups of the star transform are examined in relation to that condition.

The exposition of the material in this chapter closely follows the approach taken by Ambartsoumian and Latifi (2021). That paper introduced a new exact closed-form inversion formula for the star transform by establishing its connection with the (classical) Radon transform. It is simpler than the inversion algorithm described by Zhao *et al.* (2014) and provides an intuitive geometric insight into the problem. The formula leads to a complete characterization of the injective configurations of the star

*Reprinted by permission from Springer Nature Customer Service Centre GmbH: Springer Nature, The Journal of Geometric Analysis, Inversion and Symmetries of the Star Transform, G. Ambartsoumian and M. J. Latifi, Copyright 2021.

https://doi.org/10.1007/s12220-021-00680-7

transform and naturally yields the necessary and sufficient condition for
the stable image reconstruction described by Zhao *et al.* (2014). A detailed
analysis of this condition enables the classification of geometric setups of
the transform according to the stability of their inversion and provides a
recipe for building star configurations with the "most stable" inversions.
As an unexpected bonus, that analysis produced a proof of a long-standing
conjecture from algebraic geometry (see Conflitti, 2006) about the zero sets
of elementary symmetric polynomials.

5.1   Definitions

Let us recall the definition of the star transform introduced in formula (2.6).

Definition 5.1. The *star transform* maps a function $f : \mathbb{R}^d \to \mathbb{R}$ to a
linear combination of its integrals along a finite number of rays emanating
from a common vertex (see Figure 5.1). More formally,

$$\mathcal{S}^c f (\mathbf{x}, \Gamma) \doteq \sum_{k=1}^{m} c_k \left(\mathcal{X}f\right)(\mathbf{x}, \boldsymbol{\gamma}_k), \quad \mathbf{x} \in \mathbb{R}^d,$$

where $\mathcal{X}f$ denotes the divergent beam transform of f, i.e.

$$\mathcal{X}f(\mathbf{x}, \boldsymbol{\gamma}_k) \doteq \int_0^\infty f(\mathbf{x} + l\boldsymbol{\gamma}_k)\, dl,$$

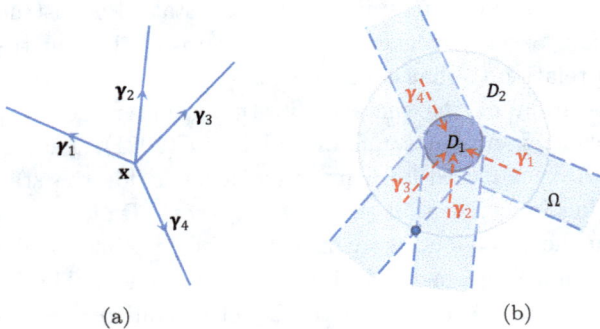

(a)   (b)

Fig. 5.1   (a) A star centered at point x with rays emanating along fixed unit vectors
$\boldsymbol{\gamma}_1, \ldots, \boldsymbol{\gamma}_4$. (b) The support of the transform integrating along the star depicted in (a).
If $\operatorname{supp} f \subseteq D_1$, then $\operatorname{supp} \mathcal{S}f \subseteq \Omega$ and $\mathcal{S}f|_{\partial D_2 \cap \Omega}$ uniquely defines $\mathcal{S}f$ in $\mathbb{R}^2 \backslash D_2$. Here,
Ω denotes the union of four semi-infinite strips containing D_1, and D_2 is the smallest
disc concentric with D_1 and containing all points on the intersections of the two strips.

$\mathbf{c} = (c_1, \ldots, c_m) \in \mathbb{R}^m$ is such that $\prod\limits_{k=1}^{m} c_k \neq 0$, and $\Gamma = [\boldsymbol{\gamma}_1, \ldots, \boldsymbol{\gamma}_m]$ is a $d \times m$ matrix with columns $\boldsymbol{\gamma}_j \in \mathbb{S}^{d-1}$, $j = 1, \ldots, m$.

Definition 5.2. We call the set

$$\bigcup_{k=1}^{m} \{\mathbf{x} + t\boldsymbol{\gamma}_k, \ t \geq 0\}$$

a *star* with a vertex at $\mathbf{x}$ and rays (or branches) along directions $\boldsymbol{\gamma}_1, \ldots, \boldsymbol{\gamma}_m$.

In this chapter, we consider a two-dimensional (2D) restriction of the star transform in the plane with fixed directions of branches, i.e. $d = 2$ and Γ is a $2 \times m$ constant matrix. As a result, the stars can be parametrized simply by the coordinates of their vertices. Since both $\mathbf{c}$ and Γ are fixed, for brevity, we will not list them in the notation of the star transform, instead writing simply $\mathcal{S}f(\mathbf{x})$. The divergent beam transform will be denoted as $\mathcal{X}_{\boldsymbol{\gamma}_k} f(\mathbf{x}) \doteq \mathcal{X}f(\mathbf{x}, \boldsymbol{\gamma}_k)$. So, in the new notations, we have the following:

$$\mathcal{S}f(\mathbf{x}) \doteq \sum_{k=1}^{m} c_k \, \mathcal{X}_{\boldsymbol{\gamma}_k} f(\mathbf{x}) = \sum_{k=1}^{m} c_k \int_0^{\infty} f(\mathbf{x} + t\boldsymbol{\gamma}_k) \, dt. \tag{5.1}$$

Definition 5.3. Let the ray directions of the star be parameterized as

$$\boldsymbol{\gamma}_j = (a_j, b_j) \in \mathbb{R}^2, \ j = 1, \ldots, m. \tag{5.2}$$

We call

$$\mathbf{a} = (a_1, \ldots, a_m) \quad \text{and} \quad \mathbf{b} = (b_1, \ldots, b_m) \tag{5.3}$$

the *aperture vectors* of that star. Since $\boldsymbol{\gamma}_j$'s are unit vectors

$$a_j^2 + b_j^2 = 1, \ j = 1, \ldots, m. \tag{5.4}$$

Every star transform corresponds to an m-gon (polygon with m sides) with vertices $c_1^{-1}\boldsymbol{\gamma}_1, \ldots, c_m^{-1}\boldsymbol{\gamma}_m$. Many properties of the star transform can be characterized by geometric features of the corresponding polygon. The following special stars play a prominent role in some of our proofs.

Definition 5.4. We call a star transform *symmetric*, if $m = 2k$ for some $k \in \mathbb{N}$ and (after possible re-indexing) $\boldsymbol{\gamma}_i = -\boldsymbol{\gamma}_{k+i}$ with $c_i = c_{k+i}$ for all $i = 1, \ldots, k$. The corresponding star is also called *symmetric*.

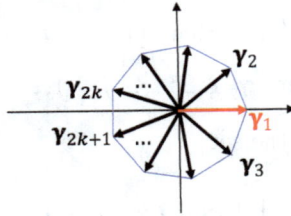

Fig. 5.2   A regular star with $m = 2k + 1$ rays and the corresponding regular m-gon.

Definition 5.5. We call a star transform *regular* if $c_1 = \cdots = c_m = 1$ and the ray directions γ_i, $i = 1, \ldots, m$ correspond to the radius vectors of the vertices of a regular m-gon (see Figure 5.2). The corresponding star is also called *regular*.

We will show that the symmetric star transforms are the only non-invertible configurations of $\mathcal{S}$, while the regular star transforms with an odd number of vertices have the most stable inversions.

For regular stars with $m = 2k + 1$ rays, we parameterize the ray directions as follows (see Figure 5.2):

$$\gamma_j = (a_j, b_j) = (\cos \alpha_j, \sin \alpha_j), \tag{5.5}$$

where

$$\alpha_1 = 0, \quad \alpha_{2j} = \frac{2\pi j}{2k+1} \quad \text{and} \quad \alpha_{2j+1} = -\frac{2\pi j}{2k+1}, \quad j = 1, \ldots, k. \tag{5.6}$$

Since $a_{2j} = a_{2j+1}$ and $b_{2j} = -b_{2j+1}$ for $j = 1, \ldots, k$, the aperture vectors of the regular star with $2k + 1$ rays can be expressed as

$$\mathbf{a} = (1, a_2, a_2, \ldots, a_{2k}, a_{2k}), \quad \mathbf{b} = (0, b_2, -b_2, \ldots, b_{2k}, -b_{2k}). \tag{5.7}$$

5.2   An Exact Inversion of the Star Transform

Let us assume that supp $f \subseteq D_1$, where D_1 is an open disc of radius r_1 centered at the origin. Then, $\mathcal{S}f$ is supported in an unbounded domain Ω, comprising a union of four semi-infinite strips containing D_1 (see Figure 5.1). At the same time, there exists a closed disc $\overline{D_2}$ of some finite radius $r_2 > r_1$ centered at the origin such that $\mathcal{S}f$ is constant along the directions of rays γ_i inside the corresponding strips of $\mathbb{R}^2 \backslash \overline{D_2}$. In other words, the restriction of $\mathcal{S}f$ to $\overline{D_2}$ completely defines it everywhere else. Throughout the rest of the text, we assume that $\mathcal{S}f(x)$ is known for all $x \in \overline{D_2}$.

In this section, we formulate and prove an exact closed-form inversion formula for nonsymmetric $\mathcal{S}$ with arbitrary geometry and arbitrary nonzero weights. The intuitive idea of this inversion is the following. The Radon transform $\mathcal{R}$ of $g = \mathcal{S}f$ at (ψ, t) is the integral of the star transform data g along the line $l(\psi, t) = \{x \in \mathbb{R}^2 | \langle x, \psi \rangle = t\}$ with normal vector ψ and signed distance t from the origin. For a generic line, this gives a sum of weighted integrals of f over half planes in $\mathbb{R}^2$, with different weights on different sides of $l(\psi, t)$. By differentiating that quantity with respect to variable t, we recover the Radon data of f along $l(\psi, t)$. The inversion of $\mathcal{S}$ is then accomplished by inverting the Radon transform.

Before proving the theorem, we discuss two auxiliary statements. They are geometric in nature and help with outlining the main ideas in the proof of Theorem 5.1.

5.2.1 *Half-plane transform and its derivative*

Definition 5.6. For a unit vector $\psi \in \mathbb{R}^2$ and $t \in \mathbb{R}$ define the *half-plane transform* of f as

$$F_\psi(t) \doteq \int_{\langle x, \psi \rangle \leq t} f \, dx. \qquad (5.8)$$

Geometrically, $F_\psi(t)$ is the integral of f over the half plane on one side of the line $l(\psi, t)$ (see Figure 5.3).

Now, let's look at the derivative of F as a function of t:

$$\frac{dF_\psi(t)}{dt} = \lim_{h \to 0} \frac{F_\psi(t + h) - F_\psi(t)}{h}.$$

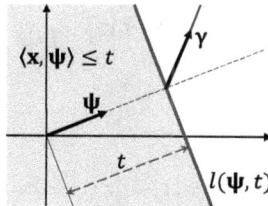

Fig. 5.3   A setup for the half plane, divergent, beam and Radon transforms.

Note that $F_{\boldsymbol{\psi}}(t+h)-F_{\boldsymbol{\psi}}(t)$ is the integral of f over the infinite strip between the lines $l(\boldsymbol{\psi},t)$ and $l(\boldsymbol{\psi},t+h)$, which plays a central role in our construction. Using Fubini's theorem and the fundamental theorem of calculus, we get

$$\frac{dF_{\boldsymbol{\psi}}(t)}{dt} = \frac{d}{dt} \int\limits_{\langle \mathbf{x}, \boldsymbol{\psi} \rangle \leq t} f \, d\mathbf{x} = \frac{d}{dt} \int_{-\infty}^{t} \mathcal{R}f(\boldsymbol{\psi},s) \, ds = \mathcal{R}f(\boldsymbol{\psi},t),$$

where $\mathcal{R}$ is the standard Radon transform. Thus, we have proved the following.

Lemma 5.1. *For function $f \in C_c(D_1)$, we have*

$$\frac{dF_{\boldsymbol{\psi}}(t)}{dt} = \mathcal{R}f(\boldsymbol{\psi},t). \tag{5.9}$$

A key step in our proof is the possibility of expressing the half-plane transform $F_{\boldsymbol{\psi}}$ of f in terms of its star transform $\mathcal{S}f$. We show this by first finding a relation between $F_{\boldsymbol{\psi}}$ and $\mathcal{X}_{\boldsymbol{\gamma}}(f)$.

If $\langle \boldsymbol{\psi}, \boldsymbol{\gamma} \rangle \neq 0$, we can integrate $\mathcal{X}_{\boldsymbol{\gamma}}(x)$ along the line $l(\boldsymbol{\psi},t)$ to get an expression in terms of $F_{\boldsymbol{\psi}}(t)$, as shown in the following.

Lemma 5.2. *Let $f \in C_c(D_1)$ and $\langle \boldsymbol{\psi}, \boldsymbol{\gamma} \rangle \neq 0$. Then,*

$$\frac{d}{dt}\mathcal{R}(\mathcal{X}_{\boldsymbol{\gamma}}f)(\boldsymbol{\psi},t) = \frac{-1}{\langle \boldsymbol{\psi}, \boldsymbol{\gamma} \rangle} \frac{dF_{\boldsymbol{\psi}}(t)}{dt}. \tag{5.10}$$

Proof. Depending on whether $\langle \boldsymbol{\psi}, \boldsymbol{\gamma} \rangle$ is positive or negative, $\mathcal{R}(\mathcal{X}_{\boldsymbol{\gamma}}f)(\boldsymbol{\psi},t)$ is integrating f with the weight $\langle \boldsymbol{\psi}, \boldsymbol{\gamma} \rangle^{-1}$ over the half plane on one or the other side of the line $l(\boldsymbol{\psi},t)$ (see Figure 5.3). This is an implication of a modified version of Fubini's theorem (e.g. see the moving sections lemma in Section A.1 of the Appendix). If $\langle \boldsymbol{\psi}, \boldsymbol{\gamma} \rangle > 0$, we have

$$\mathcal{R}(\mathcal{X}_{\boldsymbol{\gamma}}f)(\boldsymbol{\psi},t) = \frac{1}{\langle \boldsymbol{\psi}, \boldsymbol{\gamma} \rangle} \left[\int_{\mathbb{R}^2} f \, dx - F_{\boldsymbol{\psi}}(t) \right],$$

and if $\langle \boldsymbol{\psi}, \boldsymbol{\gamma} \rangle < 0$, then

$$\mathcal{R}(\mathcal{X}_{\boldsymbol{\gamma}}f)(\boldsymbol{\psi},t) = \frac{-1}{\langle \boldsymbol{\psi}, \boldsymbol{\gamma} \rangle} F_{\boldsymbol{\psi}}(t).$$

In either case, taking the derivative of both sides yields the same expression and finishes the proof. $\qquad\square$

5.2.2 Main result and some corollaries

Definition 5.7. Let $\mathcal{S}$ be the star transform defined in (5.1). We call

$$\mathcal{Z}_1 \doteq \bigcup_{j=1}^{m} \{\psi : \langle \psi, \gamma_j \rangle = 0\} \tag{5.11}$$

the *set of singular directions of Type 1 for* $\mathcal{S}$ and

$$\mathcal{Z}_2 \doteq \left\{ \psi : \sum_{j=1}^{m} c_j \prod_{i \neq j} \langle \psi, \gamma_j \rangle = 0 \right\} \tag{5.12}$$

the *set of singular directions of Type 2 for* $\mathcal{S}$. Now, define

$$w(\psi) \doteq \sum_{i=1}^{m} \frac{c_i}{\langle \psi, \gamma_i \rangle}, \ \psi \in \mathbb{S}^1 \backslash \mathcal{Z}_1; \quad q(\psi) \doteq \frac{-1}{w(\psi)}, \ \psi \in \mathbb{S}^1 \backslash (\mathcal{Z}_1 \cup \mathcal{Z}_2). \tag{5.13}$$

Theorem 5.1. *Let* $f \in C_c(D_1)$, $\mathcal{S}$ *be the star transform and* $\mathcal{R}$ *be the Radon transform of a function in* $\mathbb{R}^2$. *Then, for any* $\psi \in \mathbb{S}^1 \backslash (\mathcal{Z}_1 \cup \mathcal{Z}_2)$, *we have*

$$\mathcal{R}f(\psi, t) = q(\psi) \frac{d}{dt} \mathcal{R}(\mathcal{S}f)(\psi, t). \tag{5.14}$$

Proof. Using Lemmas 5.1 and 5.2, we get

$$\frac{d}{dt} \mathcal{R}(\mathcal{X}_\gamma f)(\psi, t) = \frac{-1}{\langle \psi, \gamma \rangle} \mathcal{R}f(\psi, t).$$

Now, we write $\mathcal{S}$ in terms of transforms $\mathcal{X}_{\gamma_i}$:

$$\frac{d}{dt} \mathcal{R}(\mathcal{S}f)(\psi, t) = \frac{d}{dt} \mathcal{R} \left(\sum_{i=1}^{m} c_i \, \mathcal{X}_{\gamma_i} f \right)(\psi, t) = \sum_{i=1}^{m} c_i \frac{d}{dt} \mathcal{R}(\mathcal{X}_{\gamma_i} f)(\psi, t)$$

$$= \sum_{i=1}^{m} \frac{-c_i}{\langle \psi, \gamma_i \rangle} \mathcal{R}f(\psi, t) = -w(\psi) \, \mathcal{R}f(\psi, t),$$

which completes the proof. $\qquad\square$

It is easy to note that if $\mathcal{Z}_1 \cup \mathcal{Z}_2$ is finite, then the singularities appearing on the right-hand side of formula (5.14) are removable.

Remark 5.1. Let $\mathcal{Z}_1 \cup \mathcal{Z}_2 = \{\zeta_i\}_{i=1}^{M}$. Then, by formula (5.14) and continuity of $\mathcal{R}f$, we have

$$\lim_{\psi \to \zeta_i} \left[q(\psi) \frac{d}{dt} \mathcal{R}(\mathcal{S}f)(\psi, t) \right] = \mathcal{R}f(\zeta_i, t) < \infty. \tag{5.15}$$

In other words, the full Radon data can be recovered from the star transform.

Corollary 5.1. *If $\mathcal{Z}_1 \cup \mathcal{Z}_2$ is finite, we can apply $\mathcal{R}^{-1}$ to recover f.*

We finish this section with statements of two special cases of Theorem 5.1 and a discussion of some pertinent relations. In the case of $m = 2$, Theorem 5.1 yields a new and simple inversion for the V-line transform $\mathcal{B}^c f$ discussed in Chapter 3.

Corollary 5.2. *An inversion formula for the V-line transform $\mathcal{B}^c = c_1 \mathcal{X}_{\gamma_1} + c_2 \mathcal{X}_{\gamma_2}$ with fixed noncollinear ray directions γ_1, γ_2 and nonzero weights c_1, c_2 is given by*

$$f = \mathcal{R}^{-1} \left[\frac{-\langle \psi, \gamma_1 \rangle \langle \psi, \gamma_2 \rangle}{c_2 \langle \psi, \gamma_1 \rangle + c_1 \langle \psi, \gamma_2 \rangle} \frac{d}{dt} \mathcal{R}(\mathcal{B}^c f)(\psi, t) \right]. \tag{5.16}$$

Another curious case of Theorem 5.1 is the following.

Corollary 5.3. *An inversion of the divergent beam transform $\mathcal{X}_\gamma$ is given by the formula*

$$f = \mathcal{R}^{-1} \left[-\langle \psi, \gamma \rangle \frac{d}{dt} \mathcal{R}(\mathcal{X}_\gamma f)(\psi, t) \right]. \tag{5.17}$$

The directional derivative $D_{-\gamma}$, given by $D_\gamma h(x) = \gamma \cdot \nabla h(x)$, is the natural inverse of $\mathcal{X}_\gamma$. Hence, for any $h \in U_\gamma = \{\mathcal{X}_\gamma(f) : f \in C_c(\mathbb{R}^2)\}$ in the range of $\mathcal{X}_\gamma$, one can write the following identity:

$$D_\gamma h(x) = \mathcal{R}^{-1} \left[\langle \psi, \gamma \rangle \frac{d}{dt} \mathcal{R} \right] h(x). \tag{5.18}$$

The identity (5.18) is not really new. It is equivalent to a known relation:

$$P(D_{e_1}, D_{e_2}) = \mathcal{R}^{-1} \left[P\left(\psi_1 \frac{d}{dt}, \psi_2 \frac{d}{dt} \right) \right] \mathcal{R}, \tag{5.19}$$

where e_i's are standard orthonormal vectors and P is a given polynomial of two variables. One can get (5.19) by repeatedly applying (5.18) and canceling $\mathcal{R}^{-1}\mathcal{R}$ in the resulting telescoping identity. The other direction of equivalence is trivial.

5.3 Injectivity of the Star Transform

In certain configurations of the star transform, the function $q(\boldsymbol{\psi})$ is not defined for any $\boldsymbol{\psi}$. For example, this happens when $m = 2$, $c_1 = c_2 = 1$ and $\boldsymbol{\gamma}_1 = -\boldsymbol{\gamma}_2 = (1,0)$. In this case, $\langle\boldsymbol{\psi},\boldsymbol{\gamma}_1\rangle + \langle\boldsymbol{\psi},\boldsymbol{\gamma}_2\rangle \equiv 0$ for any $\boldsymbol{\psi} \in \mathbb{S}^1$.

Consider a similar configuration with more rays: $m = 4$, $c_1 = \cdots = c_4 = 1$, $\boldsymbol{\gamma}_1 = -\boldsymbol{\gamma}_3$ and $\boldsymbol{\gamma}_2 = -\boldsymbol{\gamma}_4$. While it is not obvious that $q(\boldsymbol{\psi})$ is not defined at any $\boldsymbol{\psi}$, it is easy to note that in this case, the star transform is not injective since it provides less information than the classical Radon transform in 2D restricted to two directions. Interestingly enough, such configurations are the only ones for which the star transform is not injective.

Theorem 5.2. *The star transform $\mathcal{S} = \sum_{i=1}^m c_i \mathcal{X}_{\gamma_i}$ is invertible if and only if it is not symmetric.*

An important object in our inversion is the function $q(\boldsymbol{\psi})$ (or, equivalently, its reciprocal $w(\boldsymbol{\psi})$). It leads to the geometric condition that is necessary for the star transform to be noninvertible. To describe the connection between $q(\boldsymbol{\psi})$ and that geometric condition, we need to consider the *elementary symmetric polynomial* of degree $m - 1$ in m variables $y = (y_1,\ldots,y_m)$ (see Macdonald, 1998):

$$e_{m-1}(y_1,\ldots,y_m) \doteq \sum_{i=1}^m \prod_{j\neq i} y_j. \tag{5.20}$$

Using the above notation, one can rewrite formula (5.13) as

$$q(\boldsymbol{\psi}) = \frac{c_1^{-1}\ldots c_m^{-1}}{\dfrac{e_{m-1}\left(\langle\boldsymbol{\psi},c_1^{-1}\boldsymbol{\gamma}_1\rangle,\ldots,\langle\boldsymbol{\psi},c_m^{-1}\boldsymbol{\gamma}_m\rangle\right)}{\langle\boldsymbol{\psi},\boldsymbol{\gamma}_1\rangle\ldots\langle\boldsymbol{\psi},\boldsymbol{\gamma}_m\rangle}}. \tag{5.21}$$

The proof of Theorem 5.2 is based on the description of zero sets of e_{m-1} and the fact that the star transform $\mathcal{S}$ is invertible if $e_{m-1}(\langle\boldsymbol{\psi},c_1^{-1}\boldsymbol{\gamma}_1\rangle,\ldots,\langle\boldsymbol{\psi},c_m^{-1}\boldsymbol{\gamma}_m\rangle)$ is not identically zero as a function of $\boldsymbol{\psi}$.

If $\hat{\boldsymbol{\psi}} = (r,s) \in \mathbb{R}^2$ (not necessarily on $\mathbb{S}^1$) and $c \doteq c_1\ldots c_m$, we define a polynomial in two variables (r,s) as follows:

$$P_2(\hat{\boldsymbol{\psi}}) \doteq c\, e_{m-1}(\langle\hat{\boldsymbol{\psi}},c_1^{-1}\boldsymbol{\gamma}_1\rangle,\ldots,\langle\hat{\boldsymbol{\psi}},c_m^{-1}\boldsymbol{\gamma}_m\rangle). \tag{5.22}$$

Lemma 5.3. *The polynomial $P_2(r,s)$ either has finitely many zeros on $\mathbb{S}^1$ or it is identically zero.*

Proof. It is easy to note that $e_{m-1}(\mathbf{y})$ is a homogeneous polynomial of degree $d-1$, i.e. $e_{m-1}(\lambda\mathbf{y}) = \lambda^{m-1}e_{m-1}(\mathbf{y})$. Hence, $P_2(r,s)$ is also a homogeneous polynomial of degree $m-1$, which either has finitely many (projective) roots or is identically zero. □

Proof of Theorem 5.2. Assume that for a fixed choice of directions $\gamma_1, \ldots, \gamma_m$, the corresponding star transform is not invertible. By Theorem 5.1, $P_2(\boldsymbol{\psi})$ has to be zero on an infinite subset of $\mathbb{S}^1$. Therefore, by Lemma 5.3, $P_2(\boldsymbol{\psi}) \equiv 0$.

Using the notation introduced in (5.13),

$$w(\boldsymbol{\psi}) = \sum_{i=1}^{m} \frac{c_i}{\langle\boldsymbol{\psi},\boldsymbol{\gamma}_i\rangle} = \frac{P_2(\boldsymbol{\psi})}{\prod_{i=1}^{m}\langle\boldsymbol{\psi},\boldsymbol{\gamma}_i\rangle}.$$

If $w(\boldsymbol{\psi}) = 0$ away from the poles, then

$$0 = \lim_{\boldsymbol{\psi}\to\boldsymbol{\gamma}_j^\perp} \langle\boldsymbol{\psi},\boldsymbol{\gamma}_j\rangle\, w(\boldsymbol{\psi}).$$

This is c_j if there is no other $\boldsymbol{\gamma}_i$ parallel to $\boldsymbol{\gamma}_j$ and is equal to $c_j - c_{j'}$ if $\boldsymbol{\gamma}_{j'} = -\boldsymbol{\gamma}_j$. The first case cannot hold since $c_j \neq 0$, so the second must hold, which implies that $\mathcal{S}$ is symmetric. □

Theorem 5.2 immediately entails the following.

Corollary 5.4. *Any star transform with an odd number of rays is invertible.*

5.4 Singular Directions of the Star Transform

The number and location of singular directions affect the quality of numerical reconstructions. The singular directions of Type 2 correspond to "division by zero" of the processed data $\frac{d}{dt}\mathcal{R}(\mathcal{S}f)$, while those of Type 1 correspond to "multiplication by zero." Hence, it is natural to expect that singular directions of Type 2 will create instability and adversely impact the reconstruction. Our numerical experiments confirm these expectations (see Section 5.6). It is also interesting that the (totally different) algorithm for inversion of the star transform obtained by Zhao *et al.* (2014) produces a relation equivalent to the defining relation of $\mathcal{Z}_2$ in (5.12) as a necessary and sufficient condition for the instability of that algorithm (see Zhao *et al.*, 2014, p. 18, formula (51)).

While the geometric meaning of singular directions of Type 1 is obvious for any m, there is no easy interpretation of set $\mathcal{Z}_2$ for $m \geq 3$. However, the singular directions of Type 2 are more crucial to the quality of reconstruction.

In Section 5.4.1, we discuss the existence of singular directions of Type 2 in star transforms with uniform weights. In Section 5.4.2, we show the absence of Type 2 singularities in regular star transforms with an odd number of rays. The case of star transforms with nonuniform weights is discussed in Section 5.4.3.

5.4.1 *Star transforms with uniform weights*

Theorem 5.3. *Consider the star transform* $\mathcal{S} = \sum_{i=1}^{m} \mathcal{X}_{\gamma_i}$ *with uniform weights:*

(1) *If m is even, $\mathcal{S}$ must contain a singular direction of Type 2.*
(2) *If m is odd, there exist configurations of $\mathcal{S}$ that contain singular directions of Type 2 as well as configurations that do not contain them.*

Proof of Part 1. Let $m = 2k$. Then, the polynomial P_2 defined in (5.22) is an odd function on the unit circle, i.e.

$$P_2(-\psi) = -P_2(\psi), \quad \psi \in \mathbb{S}^1.$$

If $P_2 \not\equiv 0$, then there exists $\psi^* \in \mathbb{S}^1$ such that $P_2(-\psi^*) = -P_2(\psi^*) > 0$. Hence, by continuity of P_2, it has to be zero on each arc of $\mathbb{S}^1$ between ψ^* and $-\psi^*$. $\qquad\square$

Proof of Part 2a. Now, let us show that for a given odd number of rays, there exist star transforms with singularities of Type 2. In other words, for any $m = 2k + 1$, $k \in \mathbb{N}$, there exist configurations of ray directions γ_i, $i = 1, \ldots, m$, such that $P_2(\psi) = 0$ for some $\psi \in \mathbb{S}^1$, where P_2 is the polynomial defined in (5.22).

A trivial example is when a pair of rays point to opposite directions, i.e. $\gamma_i = -\gamma_j$ for some indices i and j. Then, it is easy to see that the direction normal to them is a singular direction of Type 2. A more interesting example is discussed in the following.

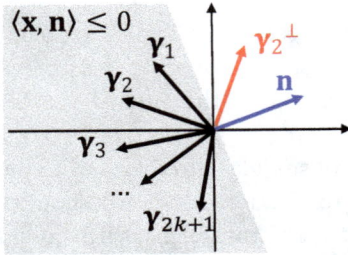

Fig. 5.4   A star with $2k + 1$ rays that has a singular direction of Type 2.

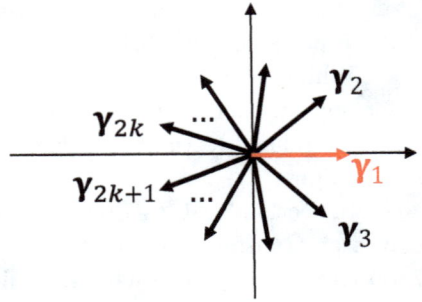

Fig. 5.5   A regular star with $2k+1$ rays without a singular direction of Type 2.

Consider an arbitrary open half plane $H = \{\mathbf{x} : \langle \mathbf{x}, \mathbf{n} \rangle < 0\}$, where $\mathbf{n}$ is the outward normal to its boundary passing through the origin. Choose a set of distinct ray directions $\{\boldsymbol{\gamma}_i\}_{i=1}^m$ so that $\boldsymbol{\gamma}_i \in H$ for all $i = 1, \ldots, m$. Without loss of generality, let us assume that $\boldsymbol{\gamma}_i$s are indexed according to the growth of their polar angle (see Figure 5.4).

Let $\boldsymbol{\psi} = \boldsymbol{\gamma}_2^{\perp}$ be oriented so that $\langle \boldsymbol{\psi}, \boldsymbol{\gamma}_1 \rangle > 0$. Then, $\langle \boldsymbol{\psi}, \boldsymbol{\gamma}_2 \rangle = 0$ and $\langle \boldsymbol{\psi}, \boldsymbol{\gamma}_i \rangle < 0$ for all $i > 2$. Estimating the terms of the sum in (5.20), it is easy to note that $\prod_{j \neq 2} \langle \boldsymbol{\psi}, \boldsymbol{\gamma}_j \rangle < 0$, while all of the other terms are zero. Thus, for $\boldsymbol{\psi} = \boldsymbol{\gamma}_2^{\perp}$, we get $P_2(\boldsymbol{\psi}) < 0$.

On the other hand, if $\boldsymbol{\psi} = \mathbf{n}$, then all the terms of the sum in (5.20) are positive and $P_2(\boldsymbol{\psi}) > 0$. By continuity of $P_2(\boldsymbol{\psi})$, there has to exist a value of $\boldsymbol{\psi}$ for which $P_2(\boldsymbol{\psi}) = 0$.   □

Proof of Part 2b. The regular star transform $\mathcal{S}$ with $m = 2k + 1$ rays does not have a singular direction of Type 2. See Section 5.4.2.   □

5.4.2   *Regular star transforms with an odd number of rays*

In this section, we show that any regular star transform with an odd number of rays (see Figure 5.5) has a stable inversion. More specifically,

Theorem 5.4.   *The regular star transform $\mathcal{S}$ with $m = 2k + 1$ rays does not have a singular direction of Type 2.*

Proof. We need to show that for a regular star with $m = 2k + 1$ rays, $P_2(\boldsymbol{\psi}) \neq 0$ for any $\boldsymbol{\psi} \in \mathbb{S}^1$, where P_2 is the polynomial defined in (5.22).

Using the parameterization $\boldsymbol{\psi}(\alpha) = (\cos \alpha, \sin \alpha)$, we denote

$$F(\alpha) = P_2(\boldsymbol{\psi}(\alpha)). \tag{5.23}$$

We show that $F(\alpha) \neq 0$ for any $\alpha \in [0, 2\pi]$. Our argument is based on the following statement, which is proved after the proof of Theorem 5.4.

Lemma 5.4. *There exists a polynomial of one variable P_1 of degree $\leq m-1$ such that*

$$F(\alpha) = P_1(\cos \alpha). \tag{5.24}$$

Due to the symmetry of $P_2(\boldsymbol{\psi})$ with respect to $\boldsymbol{\gamma}_i$s and the symmetric distributions of $\boldsymbol{\gamma}_i$s on $\mathbb{S}^1$, it follows that $F(\alpha)$ is periodic with period $2\pi/m$. Therefore, all of its Fourier coefficients F_k are zero unless k is divisible by m. Since F is a trigonometric polynomial, it is equal to its Fourier expansion, and since it is, at most, of degree $(m - 1)$, the only index in its Fourier expansion divisible by m is $k = 0$, thus F is constant. By Theorem 5.2, that constant cannot be zero, which completes the proof of Theorem 5.4. $\square$

Proof of Lemma 5.4. Recall Definition 5.3 and equation (5.7). Using our previous notations, namely $\boldsymbol{\psi}(\alpha) = (\cos \alpha, \sin \alpha) \doteq (r, s)$, we can write

$$F(\alpha) = P_2(\boldsymbol{\psi}(\alpha)) = e_{m-1}(\langle \boldsymbol{\psi}, \boldsymbol{\gamma}_1 \rangle, \ldots, \langle \boldsymbol{\psi}, \boldsymbol{\gamma}_m \rangle) = e_{m-1}(r\mathbf{a} + s\mathbf{b})$$

$$= \sum_{i=1}^{m} \prod_{j \neq i} (ra_j + sb_j)$$

$$= \prod_{j \neq 1} (ra_j + sb_j) + \left[\prod_{j \neq 2} (ra_j + sb_j) + \prod_{j \neq 3} (ra_j + sb_j) \right]$$

$$+ \cdots + \left[\prod_{j \neq 2k} (ra_j + sb_j) + \prod_{j \neq 2k+1} (ra_j + sb_j) \right].$$

It easy to note that after appropriate cancellations inside the brackets using (5.7), the last expression does not contain any odd degree of s. Using the standard trigonometric identity $s^2 = 1 - r^2$, one can eliminate the dependence on s and preserve the polynomial structure. $\square$

In the case of $m = 3$, there is a simpler proof of Theorem 5.4. That proof also shows that the transform $\mathcal{S}$ corresponding to the regular star has (in some sense) the most stable inversion. It is based on the following result from Conflitti (2006).

Theorem. *The zero set of $e_2(y_1, y_2, y_3)$ is the circular cone*

$$e_2^{-1}(0) = \left\{ \boldsymbol{y} \in \mathbb{R}^3 : \cos^2 (\boldsymbol{y}, \boldsymbol{u}) = \frac{1}{3} \right\}, \tag{5.25}$$

where $\boldsymbol{u} = (1, 1, 1)$ and $(\boldsymbol{y}, \boldsymbol{u})$ denotes the angle between vectors $\boldsymbol{y}$ and $\boldsymbol{u}$.

An alternative proof of Theorem 5.4 when $m = 3$. We need to show that $e_2(r\mathbf{a} + s\mathbf{b}) \neq 0$, where $\mathbf{a}$ and $\mathbf{b}$ are the aperture vectors of the star transform defined by (5.5), (5.6), and (5.7) as

$$\mathbf{a} = \left(1, -\frac{1}{2}, -\frac{1}{2} \right), \quad \mathbf{b} = \left(0, \frac{\sqrt{3}}{2}, -\frac{\sqrt{3}}{2} \right) \tag{5.26}$$

and $r^2 + s^2 = 1$. In light of (5.25), it is equivalent to proving that the plane T spanned by aperture vectors $\mathbf{a}$ and $\mathbf{b}$ intersects the circular cone defined in (5.25) only at the origin. Since

$$\langle \mathbf{a}, \mathbf{u} \rangle = 0 \quad \text{and} \quad \langle \mathbf{b}, \mathbf{u} \rangle = 0,$$

it is clear that $\mathbf{u} \perp T$. Hence, $T \cap e_2^{-1}(0) = \{0\}$. Moreover, the aperture vectors $\mathbf{a}$ and $\mathbf{b}$ corresponding to the regular star span the plane that is "as far as possible" from the zero set of e_2. □

Remark 5.2. The proof above shows that among the star transforms with three branches, the regular star transform has the most stable inversion.

5.4.3 *Star transforms with nonuniform weights*

The regular stars with an odd number of rays are not the only configurations of the transform $\mathcal{S}$ that do not have a singular direction of Type 2. Due to continuous dependence of $P_2(\boldsymbol{\psi})$ on α_is, small perturbations in γ_is from their symmetry positions will not introduce a singular direction of Type 2. Similarly, small perturbations in uniform weights $c_1 = \cdots = c_m = 1$ to nonuniform values will not introduce singular directions of Type 2. Moreover, it is easy to note the following.

Remark 5.3. The polynomial $P_2(\boldsymbol{\psi})$ does not change if one simultaneously replaces γ_i by $-\gamma_i$ and c_i by $-c_i$ for any i. This provides a recipe for creating star transforms with weights of mixed algebraic signs that have stable inversions.

For example, by modifying the regular star transform of three rays (defined by formula (5.26)) through replacement of γ_2 by $-\gamma_2$ and changing the sign of $c_2 = 1$, we get a stable configuration with weights $c_1 = 1$, $c_2 = -1$, and $c_3 = 1$ and aperture vectors

$$\mathbf{a} = \left(1, \frac{1}{2}, -\frac{1}{2} \right), \quad \mathbf{b} = \left(0, -\frac{\sqrt{3}}{2}, -\frac{\sqrt{3}}{2} \right).$$

In general, one can get stable setups when the absolute values of weights are close to each other.

The star transforms with weights of mixed algebraic signs satisfying the condition

$$\sum_{i=1}^{m} c_i = 0 \qquad (5.27)$$

play an important role in simultaneous reconstruction of scattering and absorption coefficients in single-scattering tomography (Florescu *et al.*, 2009, 2010; Katsevich and Krylov, 2013; Zhao *et al.*, 2014). The aims of satisfying the above condition and keeping the absolute values of weights as close to each other as possible suggest choosing the weights (up to a proportionality coefficient and re-indexing) as follows:

$$c_1 = \cdots = c_k = -\frac{1}{k}, \quad c_{k+1} = \cdots = c_{2k+1} = \frac{1}{k+1}, \quad m = 2k+1. \quad (5.28)$$

In the case of $m = 3$, this implies choosing the weights proportional to $c_1 = c_2 = 1$ and $c_3 = -2$. Such a setup was analyzed in detail by Zhao *et al.* (2014).

We finish this section with generalizations of the statements in Section 5.4.1 to the case of transforms with nonuniform weights. For the stars with an even number of rays, both the result and its proof are almost identical, as described in the following.

Theorem 5.5. *Consider the star transform $\mathcal{S} = \sum_{i=1}^{m} c_i \mathcal{X}_{\gamma_i}$. If m is even, $\mathcal{S}$ must contain a singular direction of Type 2.*

In the case of star transforms with an odd number of rays, it should be clarified in what sense we want to generalize Part 2 of Theorem 5.3. If one is allowed to choose both the weights c_i and the ray directions γ_i to claim the existence of setups with and without Type 2 singularities, then the statement has a trivial proof by choosing uniform weights and referring to Theorem 5.3. Hence, we consider the following question. Are there configurations (i.e. ray directions $\gamma_1, \ldots, \gamma_m$) with/without Type 2 singularities for a specified set of weights $c_1, \ldots, c_m$?

Theorem 5.6. *Consider a star transform $\mathcal{S} = \sum_{i=1}^{m} c_i \mathcal{X}_{\gamma_i}$, where $m = 2k+1$.*

(a) For any set of specified weights $c_1, \ldots, c_m$, there exist $\gamma_1, \ldots, \gamma_m$ such that S contains singular directions of Type 2. In other words, for any set of weights, there are configurations of S with unstable inversion.

(b) Let $m = 3$. For any set of specified weights c_1, c_2, c_3, there exist $\gamma_1, \gamma_2, \gamma_3$ such that S does not contain singular directions of Type 2. In other words, for any set of weights, there are configurations of S with a stable inversion.

Proof of (a). Existence of setups with Type 2 singularities. Both examples provided in the proof of Theorem 5.3 work here with small modifications. If $\gamma_i = -\gamma_j$ for some indices i and j, then the direction normal to them is a singular direction of Type 2.

A more interesting example is the case when all γ_is belong to the same half plane. Recall the corresponding setup from the proof of Theorem 5.3. Without loss of generality, let us assume that $c_1 c_2 > 0$, i.e. the weights c_1 and c_2 have the same sign. Since we consider the case of $m = 2k + 1$ for $k \geq 1$, the assumption above is just a matter of re-indexing the weights (if necessary).

Let $\psi_1 = \gamma_1^{\perp}$ and $\psi_2 = \gamma_2^{\perp}$ be vectors obtained by a clockwise rotation by $\pi/2$ from γ_1 and γ_2 correspondingly. Then, it is easy to verify that $P_2(\psi_1)$ and $P_2(\psi_2)$ have different signs. Hence, by continuity of P_2, there exists some ψ such that $P_2(\psi) = 0$. $\qquad\square$

Before proceeding to the proof of the second statement of Theorem 5.6, we define a new class of planes in $\mathbb{R}^3$ and discuss their properties.

Definition 5.8. We call a plane $T \subseteq \mathbb{R}^3$ "spannable by star aperture vectors" or SSAV if $T = \text{span}\{\mathbf{a}, \mathbf{b}\}$, where $\mathbf{a}, \mathbf{b} \in \mathbb{R}^3$ are aperture vectors of some star with $m = 3$ rays. In particular, this means that (a_i, b_i) are unit vectors in $\mathbb{R}^2$ for $i = 1, 2, 3$.

Not every plane (containing the origin) in $\mathbb{R}^3$ is SSAV. The following lemma gives a complete description of admissible normal vectors to an SSAV plane.

Lemma 5.5. *A vector $\mathbf{n} = (n_1, n_2, n_3) \in \mathbb{R}^3$ is normal to an SSAV plane if and only if $|n_1|, |n_2|,$ and $|n_3|$ correspond to the side lengths of some triangle. We denote the set of all such vectors by A.*

Proof. Let n be normal to a plane $T \in \mathbb{R}^3$ containing the origin. T is SSAV if and only if there exists a star with aperture vectors $\mathbf{a}, \mathbf{b}$ such that

$\langle \mathbf{n}, \mathbf{a} \rangle = \langle \mathbf{n}, \mathbf{b} \rangle = 0$. By definition of the aperture vectors, the last relation implies

$$n_1 \gamma_1 + n_2 \gamma_2 + n_3 \gamma_3 = 0,$$

where γ_1, γ_2, and γ_3 are the ray directions of the corresponding star. Since γ_is are unit vectors, the statement of the lemma follows from the above equality interpreting it as a sum of vectors tracing the perimeter of a triangle. $\square$

The set A of admissible normals to an SSAV plane has an interesting geometric description due to the three triangle inequalities:

$$|n_1| + |n_2| \geq |n_3|, \quad |n_2| + |n_3| \geq |n_1|, \quad |n_1| + |n_3| \geq |n_2|.$$

In the octant $n_1 > 0$, $n_2 > 0$, and $n_3 > 0$, it coincides with an infinite polyhedral cone with a vertex at the origin and infinite edges along vectors $(0, 1, 1)$, $(1, 0, 1)$, and $(1, 1, 0)$. We denote the portion of A in this first octant by A^+. In all the other octants, A coincides with a similar tetrahedron with the edges along appropriately signed versions of the same vectors.

Formula (5.25) shows that every plane with a normal vector $\mathbf{n} \in A^+$ intersects $e_2^{-1}(0)$ only at the origin. Moreover, among the SSAV planes, those with a normal vector in A^+ are the only ones that intersect $e_2^{-1}(0)$ only at the origin. Hence, the set A^+ contains normals to the SSAV planes that correspond (through the spanning aperture vectors) to the star transforms with uniform weights $c_1 = c_2 = c_3 = 1$ and a stable inversion, i.e. without a singular direction of Type 2.

Let

$$W \doteq \operatorname{diag}(c_1, c_2, c_3)$$

denote the diagonal matrix of weights.

Lemma 5.6. *For any set of specified weights $c_1, c_2, c_3 > 0$, there exist SSAV planes T_1, T_2 such that $T_2 = W^{-1} T_1$ and $T_i \cap e_2^{-1}(0) = \{0\}$ for $i = 1, 2$.*

Proof. Let $\mathbf{n}^{(1)}, \mathbf{n}^{(2)}$ denote some vectors normal to the planes T_1 and T_2 correspondingly. The proof will follow immediately if we show that for any $c_1, c_2, c_3 > 0$, there exist $\mathbf{n}^{(1)}, \mathbf{n}^{(2)} \in A^+$ such that $\mathbf{n}^{(2)} = W \mathbf{n}^{(1)}$.

We first note that the problem is invariant with respect to multiplication of all weights $c_1, c_2, c_3 > 0$ by a positive constant as well as their permutation. Hence, without loss of generality, we assume that $0 < c_1 \leq c_2 = 1 \leq c_3$.

A choice of the desired pair of vectors $\mathbf{n}^{(1)}, \mathbf{n}^{(2)}$ is given by

$$\mathbf{n}^{(1)} = (1, 1, 1/c_3) \in A^+, \quad \mathbf{n}^{(2)} = W\mathbf{n}^{(1)} = (c_1, 1, 1) \in A^+,$$

which completes the proof. $\qquad\qquad\qquad\qquad\qquad\qquad\qquad\qquad$ $\square$

Proof of Theorem 5.6 (b). Setups without Type 2 singularities.
Assume $m = 3$ and let $c_1, c_2, c_3 \neq 0$ be specified weights. By Remark 5.3, each stable configuration of $\mathcal{S}$ with signed weights c_1, c_2, c_3 corresponds to another stable configuration of $\mathcal{S}$ with positive weights $|c_1|, |c_2|, |c_3|$ and the other way around. Hence, without loss of generality, we will prove our statement for $c_1, c_2, c_3 > 0$.

We use our knowledge of the existence of stable configurations for every star transform with uniform weights (marked by superscript $^{(2)}$ in the following proof) to show the existence of stable configurations for the star transforms with positive weights (marked by superscript $^{(1)}$ in the following proof).

By Lemma 5.6, there exist SSAV planes T_1, T_2 such that $T_2 = W^{-1}T_1$ and $T_i \cap e_2^{-1}(0) = \{0\}$ for $i = 1, 2$. Let $\mathbf{a}^{(1)}, \mathbf{b}^{(1)}$, and $\mathbf{a}^{(2)}, \mathbf{b}^{(2)}$ be some star aperture vector pairs spanning correspondingly T_1 and T_2.

To prove the theorem, it is enough to show that for ray directions $\gamma_1^{(1)}, \gamma_2^{(1)}, \gamma_3^{(1)}$ corresponding to the aperture vectors $\mathbf{a}^{(1)}, \mathbf{b}^{(1)}$, the following relation holds:

$$P_2^{(1)}(\boldsymbol{\psi}) \doteq c_1 \langle \boldsymbol{\psi}, \gamma_2^{(1)} \rangle \langle \boldsymbol{\psi}, \gamma_3^{(1)} \rangle + c_2 \langle \boldsymbol{\psi}, \gamma_1^{(1)} \rangle \langle \boldsymbol{\psi}, \gamma_3^{(1)} \rangle + c_3 \langle \boldsymbol{\psi}, \gamma_1^{(1)} \rangle \langle \boldsymbol{\psi}, \gamma_2^{(1)} \rangle \neq 0$$

for all $\boldsymbol{\psi} \in \mathbb{S}^1$.

Due to the properties of selected planes and corresponding aperture vectors,

$$P_2^{(2)}(\boldsymbol{\psi}) \doteq e_2\left(r\mathbf{a}^{(2)} + s\mathbf{b}^{(2)}\right) \neq 0 \quad \text{for all } \boldsymbol{\psi} = (r, s) \in \mathbb{S}^1.$$

At the same time, for any $\boldsymbol{\psi} = (r, s) \in \mathbb{S}^1$, there exists $\hat{\boldsymbol{\psi}} = (\hat{r}, \hat{s}) \in \mathbb{S}^1$ such that

$$P_2^{(1)}(\boldsymbol{\psi}) = c\, e_2\left(W^{-1}(r\mathbf{a}^{(1)} + s\mathbf{b}^{(1)})\right) = k\, e_2\left(\hat{r}\mathbf{a}^{(2)} + \hat{s}\mathbf{b}^{(2)}\right) = k\, P_2^{(2)}(\hat{\boldsymbol{\psi}}) \neq 0,$$

where $c = c_1 c_2 c_3$ and k is some positive constant. In the second equality above, we use the fact that $W^{-1}(r\mathbf{a}^{(1)} + s\mathbf{b}^{(1)}) \in T_2$ and the homogeneity of e_2. $\qquad\qquad\qquad\qquad\qquad\qquad\qquad\qquad\qquad\qquad$ $\square$

The generalization of Theorem 5.6(b) to the case of $m = 2k + 1 > 3$ is more complicated, and it is still an open problem.

5.5   Proof of a Conjecture from Algebraic Geometry

In the paper by Conflitti (2006) published in 2006, the author formulated
the following conjecture about the zero sets of elementary symmetric polynomials (recall formula (5.20)):

Conjecture 5.1. *If r is even, then $e_r^{-1}(0)$ contains no real vector subspace
of dimension r.*

Furthermore, it is stated therein that one of the extreme cases is "the case
$e_{m-1}(y_1, \ldots, y_m)$, $m \equiv 1 \pmod 2$, which becomes a task quite hard to
tackle."

Our proof of Theorem 5.4 includes a proof of the aforementioned
extreme case as follows.

Theorem 5.7. *Let $m = 2k + 1$ for some $k \in \mathbb{N}$. Then, $e_{m-1}^{-1}(0)$ contains
no real vector subspace of dimension $m - 1$.*

Proof. Indeed, let $\mathbf{a}, \mathbf{b} \in \mathbb{R}^m$ be the (linearly independent) aperture
vectors of the regular star with $m = 2k + 1$ rays. It was shown that
$e_{m-1}(r\mathbf{a} + s\mathbf{b}) = P_2(\psi) \neq 0$ for any $\psi = (r, s) \in \mathbb{S}^1$. The set $C \doteq
\{r\mathbf{a} + s\mathbf{b} : (r, s) \in \mathbb{S}^1\}$ represents a closed contour around the origin in the
2D subspace $T \subset \mathbb{R}^m$ spanned by the aperture vectors $\mathbf{a}$ and $\mathbf{b}$. Due to the
homogeneity of $e_{m-1}(\mathbf{y})$, the fact that it has no zeros on C implies that
$e_{m-1}^{-1}(0) \bigcap T = \{0\}$. Since $\dim(T) = 2$, the zero set $e_{m-1}^{-1}(0)$ cannot contain
a real vector subspace of dimension $m - 1$.   □

5.6   Numerical Simulations

In this section, we present some examples of numerical reconstructions of a
simple phantom (see Figure 5.6(a)) from its star transforms of various configurations. The reconstruction algorithm is based on Theorem 5.1 and the
filtered backprojection algorithm of the Radon transform. The resolution
of all images is 300×300 pixels. The weights $c_i = 1$ for all i.

In all the following simulations, we have generated the star transform $\mathcal{S}f$
of f depicted in Figure 5.6(a) by numerically computing the divergent beam
transforms $\mathcal{X}_{\gamma_j} f$ along given directions γ_j and adding them up. Since all
$\mathcal{X}_{\gamma_j} f$s and $\mathcal{S}f$ have unbounded support (even for compactly supported f),
one has to truncate the data. In other words, numerically, $\mathcal{S}f$ is represented

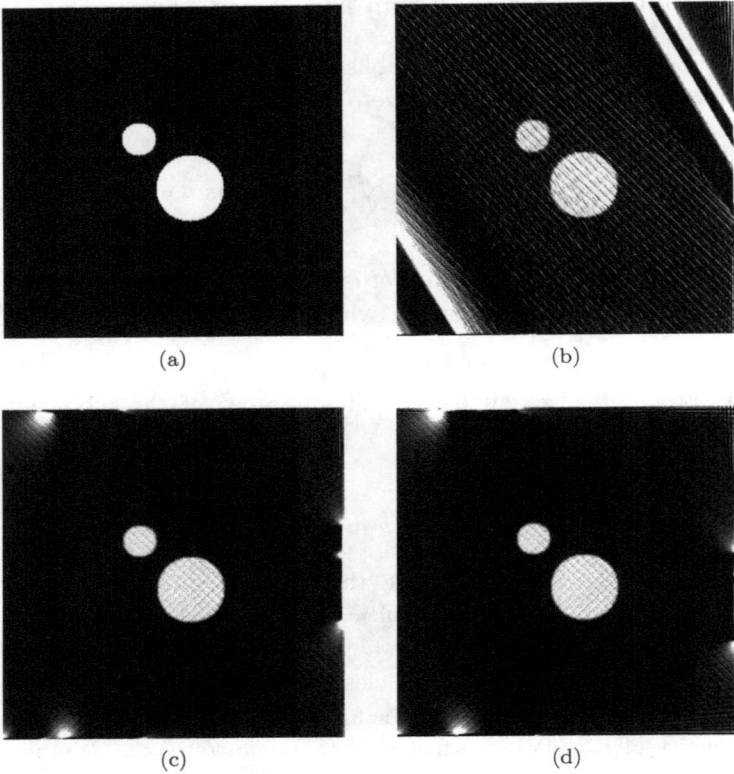

Fig. 5.6 (a) A phantom composed of two discs used in all reconstructions.
(b) Reconstruction from S with two branches: $\gamma_1 = (-1, 0)$, $\gamma_2 = (\cos(\pi/3), \sin(\pi/3))$.
(c) Reconstruction from regular S with three branches: $\gamma_1 = (-1, 0)$, $\gamma_2 = (\cos(\pi/3), \sin(-\pi/3))$, $\gamma_3 = (\cos(\pi/3), \sin(\pi/3))$. (d) Reconstruction from S with three branches: $\gamma_1 = (\cos(19\pi/20), \sin(19\pi/20))$, $\gamma_2 = (\cos(\pi/3), \sin(-\pi/3))$, $\gamma_3 = (\cos(\pi/3), \sin(\pi/3))$. *Implementation of these numerical simulations is courtesy of Mohammad J. Latifi.*

by a matrix A in which each entry $a_{i,j}$ corresponds to $Sf(x_{i,j})$ with vertices $x_{i,j}$ uniformly sampled inside the square $[-1, 1] \times [-1, 1]$, while the "true" Sf has unbounded support. Such truncation creates numerical errors when computing the Radon transform of $\mathcal{R}(Sf)(\psi, t)$ along the lines that pass through the truncated "tail" of Sf. In the reconstructed images, these errors appear in the form of artifacts at the edges of the unit square. Moreover, in unstable geometric configurations, those errors get amplified along the lines

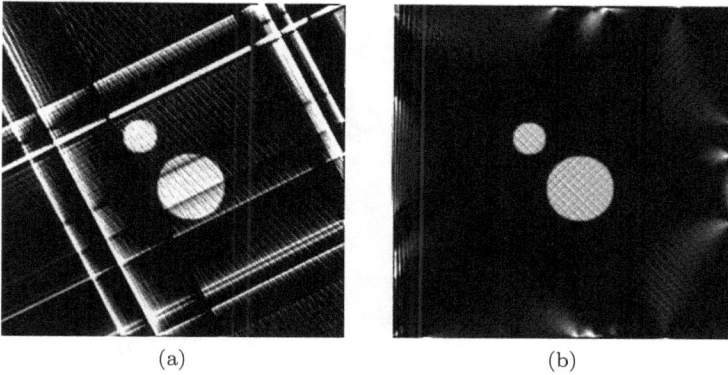

(a)   (b)

Fig. 5.7   (a) Reconstruction from $\mathcal{S}$ with three branches: $\boldsymbol{\gamma}_1 = (1,0)$, $\boldsymbol{\gamma}_2 = (0,1)$, $\boldsymbol{\gamma}_3 = (\cos(3\pi/4), \sin(3\pi/4))$. (b) Reconstruction from regular $\mathcal{S}$ with five branches: $\boldsymbol{\gamma}_1 = (1,0)$, $\boldsymbol{\gamma}_2 = (\cos(2\pi/5), \sin(2\pi/5))$, $\boldsymbol{\gamma}_3 = (\cos(4\pi/5), \sin(4\pi/5))$, $\boldsymbol{\gamma}_4 = (\cos(6\pi/5), \sin(6\pi/5))$, $\boldsymbol{\gamma}_5 = (\cos(8\pi/5), \sin(8\pi/5))$. *Implementation of these numerical simulations is courtesy of Mohammad J. Latifi.*

$l(\boldsymbol{\psi}, t)$ with the normal unit vector $\boldsymbol{\psi}$ corresponding to singular directions of Type 2 for $\mathcal{S}$ due to the presence of $q(\boldsymbol{\psi})$ in the inversion formula (5.14).

It is easy to note that in the setup corresponding to an even number of rays (see Figure 5.6(b)), the reconstruction has severe artifacts propagating along the lines with normals in the direction of Type 2 singularities. Namely, the directions of Type 2 singularities are $\boldsymbol{\psi}_1 = (\cos(\pi/6), \sin(\pi/6))$ and $\boldsymbol{\psi}_2 = (\cos(7\pi/6), \sin(7\pi/6))$, and the severe artifacts are pronounced along the lines parallel to the vector $(\cos(2\pi/3), \sin(2\pi/3))$. Similar issues can be observed in Figure 5.7(a), corresponding to the configuration with an odd number of rays and Type 2 singularities described in the proof of Part 2a of Theorem 5.3.

The setups with an odd number of rays that do not have singular directions of Type 2 are of much better quality and do not have such severe artifacts (see Figures 5.6(c) and (d) and Figure 5.7(b)). The milder artifacts close to the boundary of the unit square are almost entirely outside of the support of the image function (i.e. outside of the unit disc) and could have been cleared after the reconstruction, but we kept them for completeness of the experimental results.

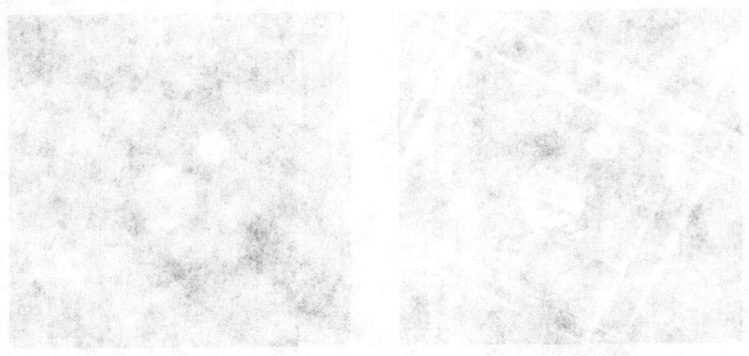

Chapter 6

Transforms on Vector Fields

All mathematical models considered in the book up to this point were based on the radiative transport equation and the solution of its single-scattered approximation discussed in Chapter 1. Those setups were based on various simplifying assumptions, relaxing which may lead to more accurate, albeit more complex, models. In particular, the authors of the first paper on SSOT, Florescu et al. (2009), argued that "since electromagnetic waves in random media are, in general, polarized, exploring the effects of polarization within the framework of the generalized vector radiative transport equation may provide additional information about the inhomogeneities in the system" (e.g. see Fernández et al., 1993; Fernández, 1999; Ishimaru, 2017).

As a step in that direction, in this chapter, we consider integral geometric problems corresponding to transforms defined on vector fields and integrating along V-lines. The reconstruction of vector fields (or, more generally, tensor fields) from generalized Radon transforms are classical problems, which are well studied in the case of smooth integration trajectories (e.g. see Sharafutdinov, 1994; Paternain et al., 2014, and the references in the next section), as well as for the cases when the trajectories reflect from (or break on) the boundaries of the image domain (e.g. see Eskin, 2004; Ilmavirta and Paternain, 2018). However, to the best of our knowledge, the recent paper by Ambartsoumian et al. (2020) is the only work dealing with the case of transforms integrating vector fields along trajectories with "vertices" inside the image domain. The exposition of the material in this chapter closely follows that article.

6.1   Introduction

Let us start with definitions of some standard generalized Radon transforms defined on tensor fields (Sharafutdinov, 1994). Assume $\mathbf{f}$ is a tensor field of rank m in $\mathbb{R}^d$ with continuous, compactly supported components.

Definition 6.1. For an integer $k \geq 0$, the *kth integral moment* of $\mathbf{f}$ (also known as the *momentum ray transform* of $\mathbf{f}$) is defined as

$$I^k \mathbf{f}(\mathbf{x}, \boldsymbol{\xi}) \doteq \int_{\mathbb{R}} t^k \, f_{i_1 \ldots i_m}(\mathbf{x} + t\boldsymbol{\xi}) \, \xi^{i_1} \ldots \xi^{i_m} \, dt.$$

The expression inside the integral uses the Einstein summation convention.

Definition 6.2. The operator $L \doteq I^0$ is called *longitudinal ray transform* (or simply *ray transform*). If $\mathbf{f}$ is a vector field, $L\mathbf{f}$ is also known as the *Doppler transform* of $\mathbf{f}$.

Definition 6.3. The *transverse ray transform* of $\mathbf{f}$ is defined as

$$T\mathbf{f}(\mathbf{x}, \boldsymbol{\xi}, \boldsymbol{\eta}) \doteq \int_{\mathbb{R}} f_{i_1 \ldots i_m}(\mathbf{x} + t\boldsymbol{\xi}) \, \eta^{i_1} \ldots \eta^{i_m} \, dt,$$

where $\boldsymbol{\eta} \perp \boldsymbol{\xi}$.

A typical problem in tensor tomography is the following. What information about a tensor field can be recovered from its ray transform? It is easy to see that in the case of a scalar function f (i.e. $m = 0$), the ray transform Lf defined above corresponds to the classical X-ray transform of f. Thus, f can be recovered uniquely from the knowledge of Lf (e.g. see Natterer and Wübbeling, 2001).

For $m \geq 1$, this transform has a nontrivial kernel, which makes the full recovery of a tensor field impossible when using only the ray transform data. In this case, only the solenoidal part of a tensor field can be recovered. The latter problem has been studied in various settings by multiple authors (e.g. see Denisjuk, 2006; Katsevich and Schuster, 2013; Krishnan and Mishra, 2018; Monard, 2016; Palamodov, 2009; Paternain *et al.*, 2013; Schuster, 2000; Sharafutdinov, 1994, 2007; Sparr *et al.*, 1995, and the references therein).

The noninjectivity of the ray transform raises a natural question: What kind of additional data is needed for full recovery of the tensor fields? The possibility of unique reconstruction of a symmetric m-tensor field $\mathbf{f}$ in $\mathbb{R}^d$ from its first $m + 1$ integral moment transforms was proved by Sharafutdinov (1986). Various other interesting results about the invertibilty of these generalized Radon transforms and their properties have been discussed in recent articles (Abhishek and Mishra, 2019; Kim and Wongsason, 2020; Krishnan *et al.*, 2019a,b, 2020; Mishra, 2020; Mishra and Sahoo, 2021).

Another approach to reconstruct the full tensor field is to work with the transverse ray transform (TRT) instead of (or in addition to) the longitudinal ray transform (LRT). If $d = 2$, TRT and LRT provide equivalent information up to a linear transformation of the tensor field. In particular, TRT also has a nontrivial kernel, making the full recovery only from TRT impossible. However, one can combine the data from both transforms (TRT and LRT) in two dimensions (2D) to recover a vector field completely (Derevtsov and Pickalov, 2011). For the recovery of tensor fields, one needs to work with mixed ray transforms, which are natural generalizations of TRT and LRT, e.g. see Derevtsov and Svetov (2015) and de Hoop *et al.* (2019). In contrast to the 2D case, when $d \geq 3$, it is known that a symmetric m-tensor field is completely determined by its TRT, see Abhishek (2020), Derevtsov and Svetov (2015), Holman (2013), Novikov and Sharafutdinov (2007), and Sharafutdinov (2008). In addition to these injectivity results, various authors have derived reconstruction methods using TRT in different settings ($d \geq 3$) (e.g. see Desai and Lionheart, 2016; Griesmaier *et al.*, 2018; Krishnan *et al.*, 2021; Wongsason, 2018, and reference therein).

Numerous other excellent works have been dedicated to the study of range conditions, support theorems, microlocal analysis, and other properties of the aforementioned generalized Radon transforms on tensor fields (e.g. see Kazantsev and Bukhgeim, 2004; Krishnan and Sharafutdinov, 2021; Krishnan and Stefanov, 2009; Ramaseshan, 2004; Sadiq *et al.*, 2016; Sadiq and Tamasan, 2015; Sharafutdinov, 2016; Stefanov and Uhlmann, 2004, 2008; Stefanov *et al.*, 2018; Vertgeim, 2000).

In this chapter, we consider a full reconstruction of a vector field in $\mathbb{R}^2$ using a new set of integral transforms (see Tables 6.1 and 6.2). These operators are analogous to the ray transforms discussed above but integrate along V-lines instead of straight lines.

Table 6.1   A list of integral operators discussed in this chapter.

Symbol	Name	Definition
$\mathcal{L}$	Longitudinal V-line transform	6.4
$\mathcal{T}$	Transverse V-line transform	6.5
$\mathcal{I}$	First moment longitudinal V-line transform	6.7
$\mathcal{J}$	First moment transverse V-line transform	6.8
$\mathcal{S}$	Vector-valued star transform	6.9

Table 6.2   A list of reconstructions provided in this chapter.

Reconstruction of **f** from:	Theorem
Knowledge of $\mathcal{L}\mathbf{f}$ and $\mathcal{T}\mathbf{f}$	6.4, 6.5
Knowledge of $\mathcal{L}\mathbf{f}$ and $\mathcal{I}\mathbf{f}$	6.6
Knowledge of $\mathcal{T}\mathbf{f}$ and $\mathcal{J}\mathbf{f}$	6.7
Knowledge of $\mathcal{S}\mathbf{f}$	6.8

6.2   Definitions and Notations

In this section, we introduce the notations and define the operators used in the chapter.

We denote by $\mathbf{x} \cdot \mathbf{y}$ the usual dot product between vectors $\mathbf{x}$ and $\mathbf{y}$. For a scalar function $V(x_1, x_2)$ and a vector field $\mathbf{f} = (f_1, f_2)$, we use the notations

$$\nabla V \doteq \left(\frac{\partial V}{\partial x_1}, \frac{\partial V}{\partial x_2} \right), \quad \operatorname{div} \mathbf{f} \doteq \frac{\partial f_1}{\partial x_1} + \frac{\partial f_2}{\partial x_2}, \quad \text{and} \quad \operatorname{curl} \mathbf{f} \doteq \frac{\partial f_2}{\partial x_1} - \frac{\partial f_1}{\partial x_2}.$$
(6.1)

The operators ∇ and div are the classical gradient and divergence operators, respectively. The operator curl defined above is essentially the exterior derivative on 2D manifolds (e.g. see Lee, 2012, Chapter 14)). However, it is customary to call that operator curl in 2D (e.g. see Derevtsov and Pickalov, 2011; Derevtsov and Svetov, 2015; Sparr *et al.*, 1995).

Let $\mathbf{u}$ and $\mathbf{v}$ be two linearly independent unit vectors in $\mathbb{R}^2$. For $\mathbf{x} \in \mathbb{R}^2$, the rays emanating from $\mathbf{x}$ in directions $\mathbf{u}$ and $\mathbf{v}$ are denoted by $L_\mathbf{u}(\mathbf{x})$ and $L_\mathbf{v}(\mathbf{x})$, respectively, i.e.

$$L_\mathbf{u}(\mathbf{x}) = \{\mathbf{x} + t\mathbf{u} : 0 \le t < \infty\} \quad \text{and} \quad L_\mathbf{v}(\mathbf{x}) = \{\mathbf{x} + t\mathbf{v} : 0 \le t < \infty\}.$$

A V-line with vertex $\mathbf{x}$ is the union of rays $L_\mathbf{u}(\mathbf{x})$ and $L_\mathbf{v}(\mathbf{x})$. In this chapter, we assume that $\mathbf{u}$ and $\mathbf{v}$ are fixed, i.e. all V-lines have the same ray directions

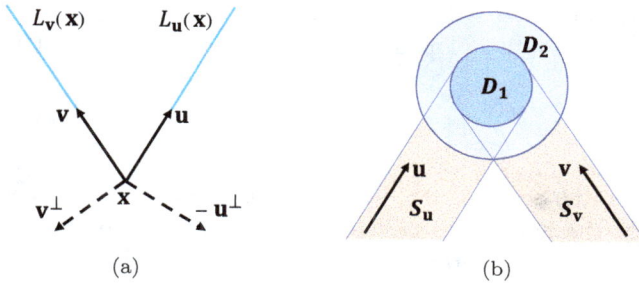

Fig. 6.1   (a) V-line with vertex at $\mathbf{x}$, ray directions $\mathbf{u}, \mathbf{v}$, and outward normals $-\mathbf{u}^{\perp}$, $\mathbf{v}^{\perp}$. (b) A sketch of the compact support of $\mathbf{f}$ and the unbounded support of $\mathcal{L}\mathbf{f}$, $\mathcal{T}\mathbf{f}$, $\mathcal{I}\mathbf{f}$, and $\mathcal{J}\mathbf{f}$.

and can be parameterized simply by the coordinates $\mathbf{x}$ of the vertex (see Figure 6.1(a)).

Following the notations used in the previous chapters of the book, the divergent beam transform $\mathcal{X}_{\mathbf{u}}$ of function h at $\mathbf{x} \in \mathbb{R}^2$ in the direction $\mathbf{u}$ is defined as

$$\mathcal{X}_{\mathbf{u}}h(\mathbf{x}) = \int_0^\infty h(\mathbf{x} + t\mathbf{u})\, dt, \tag{6.2}$$

and the directional derivative of a function in the direction $\mathbf{u}$ is denoted by $D_{\mathbf{u}}$, i.e.

$$D_{\mathbf{u}}h = \mathbf{u} \cdot \nabla h. \tag{6.3}$$

For a compactly supported function h, we observe the inverse relations (following from the fundamental theorem of calculus) between the operators $\mathcal{X}_{\mathbf{u}}$ and $D_{\mathbf{u}}$ (and similarly between $\mathcal{X}_{\mathbf{v}}$ and $D_{\mathbf{v}}$):

$$\mathcal{X}_{\mathbf{u}}\left(D_{\mathbf{u}}h\right)(\mathbf{x}) = -h(\mathbf{x}) \quad \text{and} \quad D_{\mathbf{u}}\left(\mathcal{X}_{\mathbf{u}}h\right)(\mathbf{x}) = -h(\mathbf{x}). \tag{6.4}$$

Our goal is to recover a vector field from the knowledge of its various integral transforms, namely $\mathcal{L}$, $\mathcal{T}$, $\mathcal{I}$, $\mathcal{J}$, and the star transform $\mathcal{S}$. These transforms are defined in analogy with the corresponding ray transforms of vector fields in $\mathbb{R}^2$, substituting the straight-line trajectory of integration of the latter with a V-line or star trajectory for the former. In applications, V-lines correspond to the flight paths of particles that scatter at some point in the medium. One can imagine a particle starting from a point at infinity traveling along direction $-\mathbf{u}$ to $\mathbf{x}$, where scattering happens, after which the particle goes to infinity in the direction of $\mathbf{v}$. This discussion motivates the following:

Definition 6.4. Let $\mathbf{f} = (f_1, f_2)$ be a vector field in $\mathbb{R}^2$ with components $f_i \in C_c^2(\mathbb{R}^2)$ for $i = 1, 2$. The *longitudinal V-line transform* of $\mathbf{f}$ is defined as

$$\mathcal{L}_{\mathbf{u},\mathbf{v}}\,\mathbf{f} \doteq -\mathcal{X}_{\mathbf{u}}\,(\mathbf{f}\cdot\mathbf{u}) + \mathcal{X}_{\mathbf{v}}\,(\mathbf{f}\cdot\mathbf{v}). \tag{6.5}$$

To define the second integral transform of interest, we need to make a choice for the normal unit vector corresponding to each branch of the V-line. We define the vector $\perp$ operation by $(x_1, x_2)^\perp = (-x_2, x_1)$.

Definition 6.5. Let $\mathbf{f} = (f_1, f_2)$ be a vector field in $\mathbb{R}^2$ with components $f_i \in C_c^2(\mathbb{R}^2)$ for $i = 1, 2$. The *transverse V-line transform* of $\mathbf{f}$ is defined as

$$\mathcal{T}_{\mathbf{u},\mathbf{v}}\,\mathbf{f} \doteq -\mathcal{X}_{\mathbf{u}}\,(\mathbf{f}\cdot\mathbf{u}^\perp) + \mathcal{X}_{\mathbf{v}}\,(\mathbf{f}\cdot\mathbf{v}^\perp). \tag{6.6}$$

The orientation of normal vectors is chosen toward the same side of the path of the scattering particle. Hence, in the definition above, the inner product of the unknown vector field is taken with the outward unit normal of the V-line at each point (see Figure 6.1(a)).

Definition 6.6. The *first moment divergent beam transform* of a function h in the direction $\mathbf{u}$ is defined as follows:

$$\mathcal{X}_{\mathbf{u}}^1 h(\mathbf{x}) = \int_0^\infty h(\mathbf{x} + t\mathbf{u})\,t\,dt.$$

Similarly, $\mathcal{X}_{\mathbf{v}}^1 h(\mathbf{x}) = \int_0^\infty h(\mathbf{x} + t\mathbf{v})\,t\,dt$.

Definition 6.7. Let $\mathbf{f} = (f_1, f_2)$ be a vector field in $\mathbb{R}^2$ with components $f_i \in C_c^2(\mathbb{R}^2)$ for $i = 1, 2$. The *first moment longitudinal V-line transform* of $\mathbf{f}$ is defined as

$$\mathcal{I}_{\mathbf{u},\mathbf{v}}\,\mathbf{f}\,(\mathbf{x}) = -\mathcal{X}_{\mathbf{u}}^1(\mathbf{f}\cdot\mathbf{u}) + \mathcal{X}_{\mathbf{v}}^1(\mathbf{f}\cdot\mathbf{v}). \tag{6.7}$$

Definition 6.8. Let $\mathbf{f} = (f_1, f_2)$ be a vector field in $\mathbb{R}^2$ with components $f_i \in C_c^2(\mathbb{R}^2)$ for $i = 1, 2$. The *first moment transverse V-line transform* of $\mathbf{f}$ is defined as

$$\mathcal{J}_{\mathbf{u},\mathbf{v}}\,\mathbf{f}\,(\mathbf{x}) = -\mathcal{X}_{\mathbf{u}}^1\,(\mathbf{f}\cdot\mathbf{u}^\perp) + \mathcal{X}_{\mathbf{v}}^1\,(\mathbf{f}\cdot\mathbf{v}^\perp). \tag{6.8}$$

Remark 6.1. It is easy to verify that $\mathcal{T}_{\mathbf{u},\mathbf{v}}\mathbf{f} = -\mathcal{L}_{\mathbf{u},\mathbf{v}}\mathbf{f}^\perp$ and $\mathcal{J}_{\mathbf{u},\mathbf{v}}\mathbf{f} = -\mathcal{I}_{\mathbf{u},\mathbf{v}}\mathbf{f}^\perp$.

Remark 6.2. Throughout the chapter, we assume that the linearly independent unit vectors **u** and **v** are fixed. Hence, to simplify the notations, we drop the indices **u**, **v** and refer to $\mathcal{T}_{\mathbf{u},\mathbf{v}}$, $\mathcal{L}_{\mathbf{u},\mathbf{v}}$, $\mathcal{I}_{\mathbf{u},\mathbf{v}}$, and $\mathcal{J}_{\mathbf{u},\mathbf{v}}$ simply as $\mathcal{T}$, $\mathcal{L}$, $\mathcal{I}$, and $\mathcal{J}$.

The support of f and related restrictions of its transforms: Let us assume that $\operatorname{supp} \mathbf{f} \subseteq D_1$, where D_1 is an open disc of radius r_1 centered at the origin. Let D_2 be the smallest disc (of some finite radius $r_2 > r_1$) centered at the origin such that only one ray of any V-line with a vertex outside of D_2 intersects D_1 (see Figure 6.1(b)). Then, $\mathcal{L}\mathbf{f}$, $\mathcal{T}\mathbf{f}$, $\mathcal{I}\mathbf{f}$, and $\mathcal{J}\mathbf{f}$ are supported inside an unbounded domain $D_2 \cup S_{\mathbf{u}} \cup S_{\mathbf{v}}$, where $S_{\mathbf{u}}$ and $S_{\mathbf{v}}$ are semi-infinite strips (outside of D_2) in the direction of **u** and **v**, respectively (see Figure 6.1(b)). It is easy to note that all four transforms, $\mathcal{L}\mathbf{f}$, $\mathcal{T}\mathbf{f}$, $\mathcal{I}\mathbf{f}$, and $\mathcal{J}\mathbf{f}$, are constant along the directions of rays **u** and **v** inside the corresponding strips $S_{\mathbf{u}}$ and $S_{\mathbf{v}}$. In other words, the restrictions of $\mathcal{L}\mathbf{f}$, $\mathcal{T}\mathbf{f}$, $\mathcal{I}\mathbf{f}$, and $\mathcal{J}\mathbf{f}$ to $\overline{D_2}$ completely define them in $\mathbb{R}^2$.

Remark 6.3. Throughout the chapter, we assume that the vector field **f** is supported in D_1 and the transforms $\mathcal{L}\mathbf{f}(\mathbf{x})$, $\mathcal{T}\mathbf{f}(\mathbf{x})$, $\mathcal{I}\mathbf{f}(\mathbf{x})$, and $\mathcal{J}\mathbf{f}(\mathbf{x})$ are known for all $\mathbf{x} \in \overline{D_2}$.

6.3 Full Field Recovery Using Longitudinal and Transverse VLT

The following two relations can be obtained by a simple calculation:

$$\Delta f_1 = \frac{\partial}{\partial x_1} \operatorname{div} \mathbf{f} - \frac{\partial}{\partial x_2} \operatorname{curl} \mathbf{f}, \tag{6.9}$$

$$\Delta f_2 = \frac{\partial}{\partial x_2} \operatorname{div} \mathbf{f} + \frac{\partial}{\partial x_1} \operatorname{curl} \mathbf{f}. \tag{6.10}$$

Therefore, the Laplacian of each component of a vector field **f** can be computed explicitly if one knows $\operatorname{div} \mathbf{f}$ and $\operatorname{curl} \mathbf{f}$. The knowledge of Laplacians of each component of **f** allows an explicit reconstruction of **f** with the help of Green's function on the disc D_1. Hence, we have the following.

Remark 6.4. To recover a compactly supported vector field **f** explicitly, one only needs to reconstruct $\operatorname{div} \mathbf{f}$ and $\operatorname{curl} \mathbf{f}$ from the integral transforms under consideration.

The following two theorems show that there is a nontrivial kernel for each of the integral transforms $\mathcal{L}$ and $\mathcal{T}$. Moreover, the theorems explicitly characterize those kernels.

Theorem 6.1. *The kernel of longitudinal V-line transform $\mathcal{L}$ is the set of all potential vector fields $\boldsymbol{f}$. In other words, if $\boldsymbol{f}$ is a vector field in $\mathbb{R}^2$ with components in $C_c^2(D_1)$, then*

$$\mathcal{L}\boldsymbol{f} \equiv 0 \quad \textit{if and only if } \boldsymbol{f} = \nabla V, \quad \textit{for some scalar function } V.$$

One can easily check that all potential vector fields $\mathbf{f} = \nabla V$ are curl-free (i.e. curl $\mathbf{f} = 0$) and vice versa. Thus, from the above theorem, we conclude that all curl-free vector fields are in the kernel of $\mathcal{L}$.

Theorem 6.2. *The kernel of transverse V-line transform $\mathcal{T}$ is the set of all divergence-free vector fields $\boldsymbol{f}$. In other words, if $\boldsymbol{f}$ is a vector field in $\mathbb{R}^2$ with components in $C_c^2(D_1)$, then*

$$\mathcal{T}\boldsymbol{f} \equiv 0 \quad \textit{if and only if } \operatorname{div} \boldsymbol{f} = 0.$$

Before moving on, we would like to recall here a crucial and well-known theorem which states that any vector field (with some boundary condition) can be decomposed uniquely into a divergence-free part and a curl-free part. The following decomposition result is true in more general settings, e.g. in arbitrary dimensions as well as for tensor fields. But for our needs, it is sufficient to consider the statement just for vector fields in $\mathbb{R}^2$.

Theorem 6.3 (Sharafutdinov (1994), Theorem 3.3.2). *Let Ω be a bounded domain in $\mathbb{R}^2$ and $\boldsymbol{f}$ be a vector field whose support is contained in Ω. Then, there exist a uniquely determined vector field $\mathbf{f}^s$ and a uniquely determined scalar function V satisfying*

$$\boldsymbol{f} = \mathbf{f}^s + \nabla V \quad \textit{with} \quad \operatorname{div} \mathbf{f}^s = 0 \quad \textit{and} \quad V|_{\partial\Omega} = 0. \quad (6.11)$$

The fields $\mathbf{f}^s$ and ∇V are known, respectively, as the *solenoidal part* (*divergence-free part*) and the *potential part* (*curl-free part*) of $\mathbf{f}$.

From Theorems 6.1 and 6.2, we see that the solenoidal part $\mathbf{f}^s$ and the potential part ∇V of $\mathbf{f}$ are contained in the kernels of $\mathcal{T}$ and $\mathcal{L}$, respectively. Hence, it is impossible to reconstruct the full vector field just from the knowledge of only one transform ($\mathcal{L}$ or $\mathcal{T}$). Also, observe from the above decomposition that

$$\operatorname{curl} \mathbf{f} = \operatorname{curl} \mathbf{f}^s \quad \text{and} \quad \operatorname{div} \mathbf{f} = \Delta V.$$

This implies that the problem of recovering $\operatorname{div} \mathbf{f}$ and $\operatorname{curl} \mathbf{f}$ is reduced to the determination of ΔV and $\operatorname{curl} \mathbf{f}^s$. Our next two theorems state that it is indeed possible to reconstruct ΔV and $\operatorname{curl} \mathbf{f}^s$, respectively, from the knowledge of $\mathcal{T}\mathbf{f}$ and $\mathcal{L}\mathbf{f}$.

Theorem 6.4. *Let $\mathbf{f}$ be a vector field in $\mathbb{R}^2$ with components in $C_c^2(D_1)$. Then, $\operatorname{curl}\mathbf{f}$ can be recovered from $\mathcal{L}\mathbf{f}$ as follows:*

$$\operatorname{curl}\mathbf{f} = \frac{1}{\det(\boldsymbol{v},\boldsymbol{u})} D_u D_v \mathcal{L}\mathbf{f}. \tag{6.12}$$

In particular, this implies that operator $\mathcal{L}$ is invertible over compactly supported divergence-free vector fields.

Theorem 6.5. *Let $\mathbf{f}$ be a vector field in $\mathbb{R}^2$ with components in $C_c^2(D_1)$. Then, $\operatorname{div}\mathbf{f}$ can be recovered from $\mathcal{T}\mathbf{f}$ as follows:*

$$\operatorname{div}\mathbf{f} = -\frac{1}{\det(\boldsymbol{v},\boldsymbol{u})} D_u D_v \mathcal{T}\mathbf{f}. \tag{6.13}$$

In particular, this implies that operator $\mathcal{T}$ is invertible over compactly supported curl-free vector fields.

Remark 6.5. The quantity appearing in the denominator of the expressions for $\operatorname{curl}\mathbf{f}$ and $\operatorname{div}\mathbf{f}$ is not zero since $\mathbf{u}$ and $\mathbf{v}$ are linearly independent. In other words,

$$\det(\mathbf{v},\mathbf{u}) = v_1 u_2 - u_1 v_2 = \mathbf{u}\cdot\mathbf{v}^\perp \neq 0.$$

In some cases, one may be interested in an unknown scalar potential V supported in D_1 while only having the measurements $\mathcal{T}\mathbf{f}$ of its gradient $\mathbf{f} = \nabla V$. Since $\operatorname{div}\mathbf{f} = \Delta V$, as a consequence of Theorem 6.5, we can recover the scalar function V explicitly by solving the following Dirichlet problem for the Poisson equation:

$$\begin{cases} \Delta V(\mathbf{x}) = -\dfrac{1}{\det(\mathbf{v},\mathbf{u})} D_u D_v \mathcal{T}\mathbf{f}(\mathbf{x}) & \text{in } D_1, \\ V(\mathbf{x}) \ = 0 & \text{on } \partial D_1. \end{cases}$$

Similarly, one may be interested in a compactly supported scalar function W when the measurements $\mathcal{L}\mathbf{f}$ are available only for $\mathbf{f} = (\nabla W)^\perp = (-\partial W/\partial x_2, \partial W/\partial x_1)$. In such cases, one may use the relation $\operatorname{curl}\mathbf{f} = \Delta W$ to get W by solving the following Dirichlet boundary value problem:

$$\begin{cases} \Delta W(\mathbf{x}) = \dfrac{1}{\det(\mathbf{v},\mathbf{u})} D_u D_v \mathcal{L}\mathbf{f}(\mathbf{x}) & \text{in } D_1, \\ W(\mathbf{x}) \ = 0 & \text{on } \partial D_1. \end{cases}$$

6.4   Proofs of Theorems 6.1, 6.2, 6.4, and 6.5

In this section, we prove all four previously stated theorems. We provide two proofs for each one of them: The first proof uses an analytic argument, while the second one presents a geometric explanation.

Proof of Theorem 6.1. For a given vector field $\mathbf{f} \in C_c^2(D_1)$, we want to show the existence of a scalar function V satisfying the following:

$$\mathcal{L}\mathbf{f} = 0 \quad \text{if and only if} \quad \mathbf{f} = \nabla V.$$

Analytic argument: Using the definition of $\mathcal{L}$ and applying directional derivatives along $\mathbf{u}$ and $\mathbf{v}$, we get

$$\mathcal{L}\mathbf{f} = -\mathcal{X}_{\mathbf{u}}\,(\mathbf{f}\cdot\mathbf{u}) + \mathcal{X}_{\mathbf{v}}\,(\mathbf{f}\cdot\mathbf{v}) = 0 \Longrightarrow$$
$$D_{\mathbf{u}}D_{\mathbf{v}}\,[-\mathcal{X}_{\mathbf{u}}\,(\mathbf{f}\cdot\mathbf{u}) + \mathcal{X}_{\mathbf{v}}\,(\mathbf{f}\cdot\mathbf{v})] = 0 \Longrightarrow$$
$$D_{\mathbf{v}}(\mathbf{f}\cdot\mathbf{u}) - D_{\mathbf{u}}(\mathbf{f}\cdot\mathbf{v}) = 0.$$

The same implications work also in the opposite direction, i.e.

$$D_{\mathbf{v}}(\mathbf{f}\cdot\mathbf{u}) - D_{\mathbf{u}}(\mathbf{f}\cdot\mathbf{v}) = 0 \quad \Longrightarrow \quad \mathcal{L}\mathbf{f} = 0.$$

To see this, consider

$$D_{\mathbf{v}}(\mathbf{f}\cdot\mathbf{u}) - D_{\mathbf{u}}(\mathbf{f}\cdot\mathbf{v}) = D_{\mathbf{u}}D_{\mathbf{v}}\,[-\mathcal{X}_{\mathbf{u}}\,(\mathbf{f}\cdot\mathbf{u}) + \mathcal{X}_{\mathbf{v}}\,(\mathbf{f}\cdot\mathbf{v})]$$
$$= -D_{\mathbf{v}}\,\{D_{\mathbf{u}}\,[\mathcal{X}_{\mathbf{u}}\,(\mathbf{f}\cdot\mathbf{u})]\} + D_{\mathbf{u}}\,\{D_{\mathbf{v}}\,[\mathcal{X}_{\mathbf{v}}\,(\mathbf{f}\cdot\mathbf{v})]\}.$$

Since $\mathcal{X}_{\mathbf{u}}\,(\mathbf{f}\cdot\mathbf{u})$ and $\mathcal{X}_{\mathbf{v}}\,(\mathbf{f}\cdot\mathbf{v})$ are constant, respectively, in the directions $\mathbf{u}$ and $\mathbf{v}$ outside of D_2, the functions $D_{\mathbf{u}}\,[\mathcal{X}_{\mathbf{u}}\,(\mathbf{f}\cdot\mathbf{u})]$ and $D_{\mathbf{v}}\,[\mathcal{X}_{\mathbf{v}}\,(\mathbf{f}\cdot\mathbf{v})]$ are supported in D_2. Hence, by applying $\mathcal{X}_{\mathbf{u}}\mathcal{X}_{\mathbf{v}}$ to the above equation and using relations (6.4), we get

$$\mathcal{X}_{\mathbf{u}}\mathcal{X}_{\mathbf{v}}\,[D_{\mathbf{v}}(\mathbf{f}\cdot\mathbf{u}) - D_{\mathbf{u}}(\mathbf{f}\cdot\mathbf{v})]$$
$$= \mathcal{X}_{\mathbf{u}}\mathcal{X}_{\mathbf{v}}\,\{-D_{\mathbf{v}}\,[D_{\mathbf{u}}\,(\mathcal{X}_{\mathbf{u}}\,(\mathbf{f}\cdot\mathbf{u}))] + D_{\mathbf{u}}\,[D_{\mathbf{v}}\,(\mathcal{X}_{\mathbf{v}}\,(\mathbf{f}\cdot\mathbf{v}))]\}$$
$$= -\mathcal{X}_{\mathbf{u}}\,(\mathbf{f}\cdot\mathbf{u}) + \mathcal{X}_{\mathbf{v}}\,(\mathbf{f}\cdot\mathbf{v}) = \mathcal{L}\mathbf{f}.$$

Therefore, $D_{\mathbf{v}}(\mathbf{f}\cdot\mathbf{u}) - D_{\mathbf{u}}(\mathbf{f}\cdot\mathbf{v}) = 0 \quad \Longrightarrow \quad \mathcal{L}\mathbf{f} = 0$.
Thus,

$$\mathcal{L}\mathbf{f} = 0 \quad \text{if and only if} \quad D_{\mathbf{v}}(\mathbf{f}\cdot\mathbf{u}) - D_{\mathbf{u}}(\mathbf{f}\cdot\mathbf{v}) = 0.$$

To complete the proof of this theorem, it suffices to show that

$$D_{\mathbf{v}}(\mathbf{f}\cdot\mathbf{u}) - D_{\mathbf{u}}(\mathbf{f}\cdot\mathbf{v}) = 0 \quad \text{if and only if} \quad \mathbf{f} = \nabla V$$

for some scalar function V. The following computations lead to the desired result:

$$D_{\mathbf{v}}(\mathbf{f} \cdot \mathbf{u}) - D_{\mathbf{u}}(\mathbf{f} \cdot \mathbf{v}) \tag{6.14}$$

$$= \left(v_1 \frac{\partial}{\partial x_1} + v_2 \frac{\partial}{\partial x_2}\right)(u_1 f_1 + u_2 f_2) - \left(u_1 \frac{\partial}{\partial x_1} + u_2 \frac{\partial}{\partial x_2}\right)(v_1 f_1 + v_2 f_2)$$

$$= v_1 u_1 \frac{\partial f_1}{\partial x_1} + v_1 u_2 \frac{\partial f_2}{\partial x_1} + v_2 u_1 \frac{\partial f_1}{\partial x_2} + v_2 u_2 \frac{\partial f_2}{\partial x_2} - v_1 u_1 \frac{\partial f_1}{\partial x_1} - v_2 u_1 \frac{\partial f_2}{\partial x_1}$$

$$- v_1 u_2 \frac{\partial f_1}{\partial x_2} - v_2 u_2 \frac{\partial f_2}{\partial x_2} = \det(\mathbf{v}, \mathbf{u}) \left(\frac{\partial f_2}{\partial x_1} - \frac{\partial f_1}{\partial x_2}\right) = \det(\mathbf{v}, \mathbf{u}) \operatorname{curl} \mathbf{f}.$$

Since $\mathbf{u}$ and $\mathbf{v}$ are linearly independent, we conclude that

$$D_{\mathbf{v}}(\mathbf{f} \cdot \mathbf{u}) - D_{\mathbf{u}}(\mathbf{f} \cdot \mathbf{v}) = 0 \quad \text{if and only if} \quad \operatorname{curl} \mathbf{f} = 0.$$

It is known that for discs $\operatorname{curl} \mathbf{f} = 0$ if and only if $\mathbf{f} = \nabla V$ for some scalar function V (for instance see Lee, 2012, Corollary 16.27). This completes the analytic proof of Theorem 6.1. $\qquad\square$

Geometric explanation
($\Longleftarrow$) Assume $\mathbf{f} = \nabla V$ for some scalar function V, thus $\operatorname{curl} \mathbf{f} = 0$. One can think of transformation $\mathcal{L}$ as the integral of the tangent component of the vector field $\mathbf{f}$ along branches of the V-lines, i.e.

$$\mathcal{L}\mathbf{f} = \int_{L_{\mathbf{u}} \cup L_{\mathbf{v}}} \mathbf{f} \cdot \boldsymbol{\tau} \, dt,$$

where $\boldsymbol{\tau}$ is the unit tangent vector of the V-line (as shown in Figure 6.2(a)).

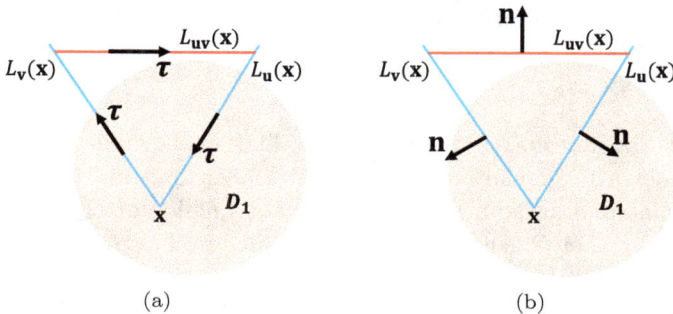

Fig. 6.2 (a) A V-line $L_{\mathbf{u}} \cup L_{\mathbf{v}}$ and an additional line segment $L_{\mathbf{uv}}$ outside of supp $\mathbf{f}$ with unit tangent vectors $\boldsymbol{\tau}$. (b) A V-line $L_{\mathbf{u}} \cup L_{\mathbf{v}}$ and an additional line segment $L_{\mathbf{uv}}$ outside of supp $\mathbf{f}$ with unit normal vectors $\mathbf{n}$.

Consider a triangular, closed contour defined by some finite intervals of the V-line and an additional "bridge" L_{uv} outside of $D_1 \supseteq \text{supp}\,\mathbf{f}$ (see Figure 6.2(a)). Let G denote the region enclosed by $L_u \cup L_v \cup L_{uv}$. Using Green's theorem and the fact that $\text{curl}\,\mathbf{f} = 0$, we get

$$\mathcal{L}\mathbf{f} = \mathcal{L}\mathbf{f} + \int_{L_{uv}} \mathbf{f} \cdot \boldsymbol{\tau}\,dt = \int_{L_u \cup L_v \cup L_{uv}} \mathbf{f} \cdot \boldsymbol{\tau}\,dt = \int_G \text{curl}\,\mathbf{f}\,ds = 0.$$

($\Longrightarrow$) The other direction of the statement in Theorem 6.1 is a consequence of Theorem 6.4.

Proof of Theorem 6.2. Recall that we want to prove that a vector field $\mathbf{f}$ is in the kernel of $\mathcal{T}$ if and only if the vector field $\mathbf{f}$ is divergence-free.

Analytic argument

Due to the following special relation between curl and divergence in $\mathbb{R}^2$:

$$\text{curl}\,\mathbf{f}^{\perp} = \text{curl}\,(-f_2, f_1) = \frac{\partial f_1}{\partial x_1} - \frac{\partial(-f_2)}{\partial x_2} = \text{div}\,\mathbf{f} \qquad (6.15)$$

and the fact that $\mathcal{T}\mathbf{f} = -\mathcal{L}\,\mathbf{f}^{\perp}$, Theorem 6.1 implies

$$\mathcal{L}\mathbf{f}^{\perp} = 0 \Longleftrightarrow \text{curl}\,\mathbf{f}^{\perp} = 0.$$

Hence,

$$\mathcal{T}\mathbf{f} = 0 \Longleftrightarrow \text{div}\,\mathbf{f} = 0,$$

which completes the proof. $\qquad\qquad\qquad\qquad\qquad\qquad\qquad\qquad\quad$ $\square$

Geometric explanation

($\Longleftarrow$) Assume $\text{div}\,\mathbf{f} = 0$. One can think of $\mathcal{T}\mathbf{f}$ as the integral of the normal component of the vector field $\mathbf{f}$ along branches of the V-lines, i.e.

$$\mathcal{T}\mathbf{f} = \int_{L_u \cup L_v} \mathbf{f} \cdot \mathbf{n}\,dt,$$

where $\mathbf{n}$ is the unit normal vector of the V-line (as shown in Figure 6.2(b)). Consider a triangular, closed contour defined by some finite intervals of the V-line and an additional "bridge" L_{uv} outside of $D_1 \supseteq \text{supp}\,\mathbf{f}$ (see Figure 6.2(b)). Let G denote the region enclosed by $L_u \cup L_v \cup L_{uv}$. We have

$$\mathcal{T}\mathbf{f} = \mathcal{T}\mathbf{f} + \int_{L_{uv}} \mathbf{f} \cdot \mathbf{n}\,dt = \int_{L_u \cup L_v \cup L_{uv}} \mathbf{f} \cdot \mathbf{n}\,dt = \int_G \text{div}\,\mathbf{f}\,ds = 0.$$

($\Longrightarrow$) The other direction of the statement in Theorem 6.2 is a consequence of Theorem 6.5.

Proof of Theorem 6.4. Analytic argument:
Recall the decomposition of a compactly supported vector field $\mathbf{f}$ presented in formula (6.11):

$$\mathbf{f} = \mathbf{f}^s + \nabla V, \quad \text{with div } \mathbf{f}^s = 0 \text{ and } V = 0 \text{ on } \partial D_1.$$

By applying $\mathcal{L}$ to this decomposition and using the fact that $\mathcal{L}(\nabla V) = 0$ (from Theorem 6.1), we get

$$\mathcal{L}\mathbf{f}(\mathbf{x}) = \mathcal{L}\mathbf{f}^s(\mathbf{x}).$$

Taking the directional derivatives $D_{\mathbf{u}}D_{\mathbf{v}}$ of the above equation and using formula (6.14), we get

$$D_{\mathbf{u}}D_{\mathbf{v}}\mathcal{L}\mathbf{f} = D_{\mathbf{u}}D_{\mathbf{v}}\left[-\mathcal{X}_{\mathbf{u}}\left(\mathbf{f}^s \cdot \mathbf{u}\right) + \mathcal{X}_{\mathbf{v}}\left(\mathbf{f}^s \cdot \mathbf{v}\right)\right] = D_{\mathbf{v}}(\mathbf{f}^s \cdot \mathbf{u}) - D_{\mathbf{u}}(\mathbf{f}^s \cdot \mathbf{v})$$
$$= \det(\mathbf{v}, \mathbf{u})\left(\frac{\partial f_2^s}{\partial x_1} - \frac{\partial f_1^s}{\partial x_2}\right).$$

Hence,

$$\frac{\partial f_2^s}{\partial x_1} - \frac{\partial f_1^s}{\partial x_2} = \frac{1}{\det(\mathbf{v}, \mathbf{u})}D_{\mathbf{u}}D_{\mathbf{v}}\mathcal{L}\mathbf{f}.$$

Finally, we observe that

$$\frac{\partial f_2^s}{\partial x_1} - \frac{\partial f_1^s}{\partial x_2} = \text{curl } \mathbf{f}^s = \text{curl } \mathbf{f}.$$

Combining the last relation with equation (6.4), we get the required expression for curl $\mathbf{f}$:

$$\text{curl } \mathbf{f} = \frac{1}{\det(\mathbf{v}, \mathbf{u})}D_{\mathbf{u}}D_{\mathbf{v}}\mathcal{L}\mathbf{f}.$$

This completes the proof of Theorem 6.4. $\qquad\square$

Geometric explanation
Consider the scalar function $h(\mathbf{x}) \doteq \mathcal{L}\mathbf{f}(\mathbf{x})$ and the following finite difference of its values at the vertices of a rhombus (refer to Figure 6.3 for visualization):

$$C\mathbf{f}(\mathbf{x}, \mathbf{y}, \mathbf{z}, \mathbf{w}) \doteq [h(\mathbf{x}) - h(\mathbf{y})] - [h(\mathbf{z}) - h(\mathbf{w})] = h(\mathbf{x}) - h(\mathbf{y}) - h(\mathbf{z}) + h(\mathbf{w}).$$

If the side length $\delta > 0$ of the rhombus tends to zero, then

$$\lim_{\delta \to 0}\left[\frac{1}{\delta^2}C\mathbf{f}(\mathbf{x}, \mathbf{y}, \mathbf{z}, \mathbf{w})\right] = D_{\mathbf{u}}D_{\mathbf{v}}h(\mathbf{x}). \tag{6.16}$$

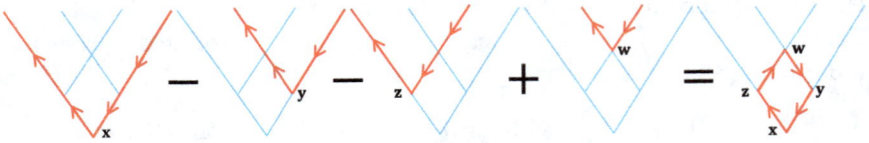

Fig. 6.3   A linear combination of $\mathcal{L}f$ at the vertices of a rhombus resulting in a contour integral of $\mathbf{f} \cdot \boldsymbol{\tau}$ along the boundary of the rhombus.

On the other hand, from Figure 6.3, it is easy to see that $C\mathbf{f}(\mathbf{x}, \mathbf{y}, \mathbf{z}, \mathbf{w})$ is the clockwise, contour integral of $\mathbf{f} \cdot \boldsymbol{\tau}$ along the boundary of the rhombus. At the same time, by the definition of curl (equivalent to the one in (6.1)) for any infinitesimal region P containing $\mathbf{x}$, we have

$$\operatorname{curl} \mathbf{f}(\mathbf{x}) = \lim_{|P| \to 0} \frac{1}{|P|} \oint_{\partial P} \mathbf{f} \cdot \boldsymbol{\tau} \, dt, \tag{6.17}$$

where the integral is taken along the contour traversed counterclockwise. Since the area of our infinitesimal rhombus is $-\delta^2 \det(\mathbf{v}, \mathbf{u})$, formulas (6.16) and (6.17) imply that

$$\operatorname{curl} \mathbf{f} = \frac{1}{\det(\mathbf{v}, \mathbf{u})} D_{\mathbf{u}} D_{\mathbf{v}} h,$$

which is what we wanted to show.

Proof of Theorem 6.5. Analytic argument:
Using the formula of Theorem 6.4 and relation (6.15) between divergence and curl, we have

$$\operatorname{curl} \mathbf{f}^{\perp} = \frac{1}{\det(\mathbf{v}, \mathbf{u})} D_{\mathbf{u}} D_{\mathbf{v}} \, \mathcal{L}\mathbf{f}^{\perp}, \tag{6.18}$$

which translates into

$$\operatorname{div} \mathbf{f} = -\frac{1}{\det(\mathbf{v}, \mathbf{u})} D_{\mathbf{u}} D_{\mathbf{v}} \, \mathcal{T}\mathbf{f}, \tag{6.19}$$

and this concludes the analytic proof of Theorem 6.5. $\square$

Geometric explanation
The argument is very similar to that of Theorem 6.4, except $h(\mathbf{x}) \doteq \mathcal{T}\mathbf{f}(\mathbf{x})$ here. Adding and subtracting the values of h as before, we obtain the outward flux of the vector field from the boundary of the infinitesimal rhombus

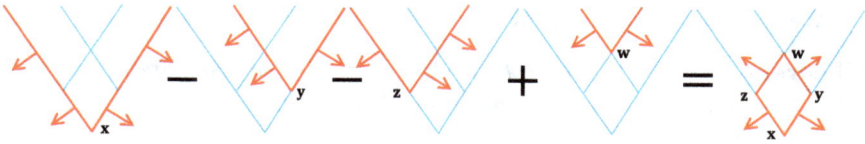

Fig. 6.4   A linear combination of $\mathcal{T}\mathbf{f}$ at the vertices of a rhombus resulting in a contour integral of $\mathbf{f} \cdot \mathbf{n}$ along the boundary of the rhombus.

(see Figure 6.4). At the same time, by the definition of divergence (equivalent to the one in (6.1)) for any infinitesimal region P containing $\mathbf{x}$, we have

$$\operatorname{div} \mathbf{f}(\mathbf{x}) = \lim_{|P| \to 0} \frac{1}{|P|} \oint_{\partial P} \mathbf{f} \cdot \mathbf{n} \, dt.$$

Hence,

$$\operatorname{div} \mathbf{f} = \frac{-1}{\det(\mathbf{v}, \mathbf{u})} D_{\mathbf{u}} D_{\mathbf{v}} h.$$

Taking $\mathbf{f} = \nabla V$, we can complete the proof using $\Delta V = \operatorname{div} \nabla V$.

6.5   Longitudinal and Transverse VLTs with Their First Moments

In this section, we show that the full vector field $\mathbf{f}$ can be recovered from the knowledge of its longitudinal V-line transform $\mathcal{L}\mathbf{f}$ and its first moment V-line transform $\mathcal{I}\mathbf{f}$ or, alternatively, from the knowledge of its transverse V-line transform $\mathcal{T}\mathbf{f}$ and its first moment V-line transform $\mathcal{J}\mathbf{f}$.

The proofs of the theorems presented in this section use the signed V-line transform $\mathcal{B}^-$ of a compactly supported scalar function h (recall formula (3.16)):

$$\mathcal{B}^- h \doteq \mathcal{X}_{\mathbf{u}} h - \mathcal{X}_{\mathbf{v}} h$$

and its explicit inversion formula (recall formula (3.25)):

$$h(\mathbf{x}) = \frac{1}{||\mathbf{v} - \mathbf{u}||} D_{\mathbf{u}} D_{\mathbf{v}} \int_0^\infty (\mathcal{B}^- h)(\mathbf{x} + \mathbf{w}t) \, dt, \qquad (6.20)$$

where

$$\mathbf{w} = \frac{\mathbf{v} - \mathbf{u}}{||\mathbf{v} - \mathbf{u}||}.$$

We can now state and prove the main results of this section.

Theorem 6.6. *Let $\mathbf{f}$ be a vector field in $\mathbb{R}^2$ with components in $C_c^2(D_1)$. Then, $\mathbf{f}$ can be recovered explicitly from $\mathcal{L}\mathbf{f}$ and $\mathcal{I}\mathbf{f}$.*

Proof. We know from Theorem 6.4 that $\operatorname{curl}\mathbf{f}$ can be expressed in terms of $\mathcal{L}\mathbf{f}$ as follows:

$$\operatorname{curl}\mathbf{f} = \frac{1}{\det(\mathbf{v},\mathbf{u})}\, D_{\mathbf{u}} D_{\mathbf{v}}\,\mathcal{L}\mathbf{f}.$$

To prove the theorem, we show that the signed V-line transform for each component of $\mathbf{f}$ can be computed explicitly in terms of $\operatorname{curl}\mathbf{f}$ and $\mathcal{I}\mathbf{f}$. Indeed,

$$\frac{\partial \mathcal{I}\mathbf{f}}{\partial x_1} = -\int_0^\infty t\left(u_1\frac{\partial f_1}{\partial x_1} + u_2\frac{\partial f_2}{\partial x_1}\right)(\mathbf{x}+t\mathbf{u})\,dt$$

$$+\int_0^\infty t\left(v_1\frac{\partial f_1}{\partial x_1} + v_2\frac{\partial f_2}{\partial x_1}\right)(\mathbf{x}+t\mathbf{v})\,dt$$

$$= -\int_0^\infty t\left(u_1\frac{\partial f_1}{\partial x_1} + u_2\frac{\partial f_1}{\partial x_2}\right)(\mathbf{x}+t\mathbf{u})\,dt$$

$$- u_2\int_0^\infty t\left(\frac{\partial f_2}{\partial x_1} - \frac{\partial f_1}{\partial x_2}\right)(\mathbf{x}+t\mathbf{u})\,dt$$

$$+\int_0^\infty t\left(v_1\frac{\partial f_1}{\partial x_1} + v_2\frac{\partial f_1}{\partial x_2}\right)(\mathbf{x}+t\mathbf{v})\,dt$$

$$+ v_2\int_0^\infty t\left(\frac{\partial f_2}{\partial x_1} - \frac{\partial f_1}{\partial x_2}\right)(\mathbf{x}+t\mathbf{v})\,dt$$

$$= -\int_0^\infty t\,\frac{d}{dt}f_1(\mathbf{x}+t\mathbf{u})\,dt$$

$$+\int_0^\infty t\,\frac{d}{dt}f_1(\mathbf{x}+t\mathbf{v})\,dt - u_2\,\mathcal{X}_{\mathbf{u}}^1(\operatorname{curl}\mathbf{f}) + v_2\,\mathcal{X}_{\mathbf{v}}^1(\operatorname{curl}\mathbf{f})$$

$$= \int_0^\infty f_1(\mathbf{x}+t\mathbf{u})\,dt - \int_0^\infty f_1(\mathbf{x}+t\mathbf{v})\,dt$$

$$- u_2\,\mathcal{X}_{\mathbf{u}}^1(\operatorname{curl}\mathbf{f}) + v_2\,\mathcal{X}_{\mathbf{v}}^1(\operatorname{curl}\mathbf{f}). \tag{6.21}$$

In other words,

$$\mathcal{X}_\mathbf{u} f_1 - \mathcal{X}_\mathbf{v} f_1 = \frac{\partial \mathcal{I}\mathbf{f}}{\partial x_1} + u_2 \, \mathcal{X}_\mathbf{u}^1(\operatorname{curl}\mathbf{f}) - v_2 \, \mathcal{X}_\mathbf{v}^1(\operatorname{curl}\mathbf{f}). \tag{6.22}$$

Differentiating $\mathcal{I}\mathbf{f}$ with respect to x_2 and proceeding with a similar calculation, we get

$$\mathcal{X}_\mathbf{u} f_2 - \mathcal{X}_\mathbf{v} f_2 = \frac{\partial \mathcal{I}\mathbf{f}}{\partial x_2} - u_1 \, \mathcal{X}_\mathbf{u}^1(\operatorname{curl}\mathbf{f}) + v_1 \, \mathcal{X}_\mathbf{v}^1(\operatorname{curl}\mathbf{f}). \tag{6.23}$$

Equations (6.22) and (6.23) express $\mathcal{B}^- f_1$ and $\mathcal{B}^- f_2$ in terms of known $\operatorname{curl}\mathbf{f}$ and $\mathcal{I}\mathbf{f}$. Therefore, we can recover f_1 and f_2 explicitly by the direct application of formula (6.20). $\square$

Theorem 6.7. *Let $\mathbf{f}$ be a vector field in $\mathbb{R}^2$ with components in $C_c^2(D_1)$. Then, $\mathbf{f}$ can be recovered explicitly from $\mathcal{T}\mathbf{f}$ and $\mathcal{J}\mathbf{f}$.*

Proof. From Theorem 6.5, we know that $\operatorname{div}\mathbf{f}$ can be expressed in terms of $\mathcal{T}\mathbf{f}$ as follows:

$$\operatorname{div}\mathbf{f} = -\frac{1}{\det(\mathbf{v}, \mathbf{u})} \, D_\mathbf{u} D_\mathbf{v} \mathcal{T}\mathbf{f}.$$

In this case, we show that the signed V-line transform of each component of $\mathbf{f}$ can be computed explicitly in terms of $\operatorname{div}\mathbf{f}$ and $\mathcal{J}\mathbf{f}$. Indeed, since $\mathcal{J}\mathbf{f} = -\mathcal{I}\mathbf{f}^\perp$, we can use (6.22) to get

$$\frac{\partial \mathcal{J}\mathbf{f}}{\partial x_1} = -\frac{\partial \mathcal{I}\mathbf{f}^\perp}{\partial x_1} = -\mathcal{X}_\mathbf{u}(\mathbf{f}^\perp)_1 + \mathcal{X}_\mathbf{v}(\mathbf{f}^\perp)_1 + u_2 \, \mathcal{X}_\mathbf{u}^1(\operatorname{curl}\mathbf{f}^\perp) - v_2 \, \mathcal{X}_\mathbf{v}^1(\operatorname{curl}\mathbf{f}^\perp)$$

$$= \mathcal{X}_\mathbf{u} f_2 - \mathcal{X}_\mathbf{v} f_2 + u_2 \, \mathcal{X}_\mathbf{u}^1(\operatorname{div}\mathbf{f}) - v_2 \, \mathcal{X}_\mathbf{v}^1(\operatorname{div}\mathbf{f}),$$

where in the last equality, we used the relations

$$(\mathbf{f}^\perp)_1 = -f_2 \quad \text{and} \quad \operatorname{curl}\mathbf{f}^\perp = \operatorname{div}\mathbf{f}.$$

Therefore, we have

$$\mathcal{X}_\mathbf{u} f_2 - \mathcal{X}_\mathbf{v} f_2 = \frac{\partial \mathcal{J}\mathbf{f}}{\partial x_1} - u_2 \, \mathcal{X}_\mathbf{u}^1(\operatorname{div}\mathbf{f}) + v_2 \, \mathcal{X}_\mathbf{v}^1(\operatorname{div}\mathbf{f}).$$

Differentiating $\mathcal{J}\mathbf{f}$ with respect to x_2 and proceeding in a similar way, we get

$$\mathcal{X}_\mathbf{u} f_1 - \mathcal{X}_\mathbf{v} f_1 = -\frac{\partial \mathcal{J}\mathbf{f}}{\partial x_2} - u_1 \, \mathcal{X}_\mathbf{u}^1(\operatorname{div}\mathbf{f}) + v_1 \, \mathcal{X}_\mathbf{v}^1(\operatorname{div}\mathbf{f}).$$

The last two relations express $\mathcal{B}^- f_1$ and $\mathcal{B}^- f_2$ in terms of known $\operatorname{div}\mathbf{f}$ and $\mathcal{J}\mathbf{f}$. Hence, we can recover f_1 and f_2 explicitly by the direct application of formula (6.20). $\square$

6.6   Recovery of the Full Vector Field from its Star Transform

In this section, we derive an inversion formula for the star transform of vector-valued functions. Our reconstruction is analogous to the inversion of the star transform of scalar functions presented in Chapter 5.

Definition 6.9. Let $\boldsymbol{\gamma}_1, \ldots, \boldsymbol{\gamma}_m$ be a distinct set of unit vectors in $\mathbb{R}^2$. The corresponding *star transform* $\mathcal{S}\mathbf{f}$ of a vector field $\mathbf{f}$ is defined by

$$\mathcal{S}\mathbf{f} \doteq \sum_{i=1}^{m} c_i \, \mathcal{X}_{\boldsymbol{\gamma}_i} \begin{bmatrix} \mathbf{f} \cdot \boldsymbol{\gamma}_i \\ \mathbf{f} \cdot \boldsymbol{\gamma}_i^{\perp} \end{bmatrix}, \tag{6.24}$$

where $c_1, \ldots, c_m$ is a set of nonzero real-valued weights.

Note that in contrast to our definition of the V-line transform, the star transform data for vector fields contain both the longitudinal and transverse components (this simplifies our discussion). Now, let $\mathcal{R}h(\boldsymbol{\psi}, s)$ denote the ordinary Radon transform of a scalar function h in $\mathbb{R}^2$ along the line normal to the unit vector $\boldsymbol{\psi}$ and at a signed distance s from the origin. Lemmas 5.1 and 5.2 provide the following identity:

$$\frac{d}{ds} \mathcal{R}(\mathcal{X}_{\boldsymbol{\gamma}_i} h)(\boldsymbol{\psi}, s) = \frac{-1}{\boldsymbol{\psi} \cdot \boldsymbol{\gamma}_i} \, \mathcal{R}h(\boldsymbol{\psi}, s). \tag{6.25}$$

Definition 6.10. Consider the vector-valued star transform $\mathcal{S}\mathbf{f}$ with branch directions $\boldsymbol{\gamma}_1, \ldots, \boldsymbol{\gamma}_m$. We call

$$\mathcal{Z}_1 \doteq \cup_{i=1}^{m} \{\boldsymbol{\psi} : \; \boldsymbol{\psi} \cdot \boldsymbol{\gamma}_i = 0\} \tag{6.26}$$

the set of singular directions of type 1 for $\mathcal{S}$.

Now, let

$$\boldsymbol{\gamma}(\boldsymbol{\psi}) \doteq -\sum_{i=1}^{m} \frac{c_i \, \boldsymbol{\gamma}_i}{\boldsymbol{\psi} \cdot \boldsymbol{\gamma}_i} \in \mathbb{R}^2, \quad \boldsymbol{\psi} \in \mathbb{S}^1 \setminus \mathcal{Z}_1. \tag{6.27}$$

We call

$$\mathcal{Z}_2 \doteq \{\boldsymbol{\psi} : \; \boldsymbol{\gamma}(\boldsymbol{\psi}) = 0\} \tag{6.28}$$

the set of singular directions of type 2 for $\mathcal{S}$.

Theorem 6.8. *Let $\boldsymbol{f}$ be a vector field in $\mathbb{R}^2$ with components in $C_c^2(D_1)$. Consider its vector-valued star transform $\boldsymbol{Sf}$ with branch directions $\boldsymbol{\gamma}_1, \ldots, \boldsymbol{\gamma}_m$, and let*

$$Q(\boldsymbol{\psi}) \doteq \begin{bmatrix} \boldsymbol{\gamma}(\boldsymbol{\psi}) \\ \boldsymbol{\gamma}(\boldsymbol{\psi})^{\perp} \end{bmatrix}^{-1} \in GL(2,\mathbb{R}), \quad \boldsymbol{\psi} \in \mathbb{S}^1 \setminus (\mathcal{Z}_1 \cup \mathcal{Z}_2). \tag{6.29}$$

Then, for any $\boldsymbol{\psi} \in \mathbb{S}^1 \setminus (\mathcal{Z}_1 \cup \mathcal{Z}_2)$ and any $s \in \mathbb{R}$, we have

$$Q(\boldsymbol{\psi}) \frac{d}{ds} \mathcal{R}(\boldsymbol{Sf})(\boldsymbol{\psi}, s) = \mathcal{R}\boldsymbol{f}(\boldsymbol{\psi}, s), \tag{6.30}$$

where $\mathcal{R}\boldsymbol{f}$ is the component-wise Radon transform of a vector field in $\mathbb{R}^2$.

Proof. From (6.25), we get

$$\frac{d}{ds}\mathcal{R}(\boldsymbol{Sf}) = \sum_{i=1}^{m} c_i \frac{d}{ds} \mathcal{R}\mathcal{X}_{\gamma_i} \begin{bmatrix} \boldsymbol{f} \cdot \boldsymbol{\gamma}_i \\ \boldsymbol{f} \cdot \boldsymbol{\gamma}_i^{\perp} \end{bmatrix} = -\sum_{i=1}^{m} \frac{c_i}{\boldsymbol{\psi} \cdot \boldsymbol{\gamma}_i} \mathcal{R} \begin{bmatrix} \boldsymbol{f} \cdot \boldsymbol{\gamma}_i \\ \boldsymbol{f} \cdot \boldsymbol{\gamma}_i^{\perp} \end{bmatrix}. \tag{6.31}$$

Using the linearity of $\mathcal{R}$ and the inner product, we simplify the last expression further to obtain

$$\frac{d}{ds}\mathcal{R}(\boldsymbol{Sf}) = \begin{bmatrix} \mathcal{R}\boldsymbol{f} \cdot \boldsymbol{\gamma}(\boldsymbol{\psi}) \\ \mathcal{R}\boldsymbol{f} \cdot \boldsymbol{\gamma}(\boldsymbol{\psi})^{\perp} \end{bmatrix} = Q(\boldsymbol{\psi})^{-1}\mathcal{R}\boldsymbol{f}. \tag{6.32}$$

Finally,

$$Q(\boldsymbol{\psi}) \frac{d}{ds} \mathcal{R}(\boldsymbol{Sf})(\boldsymbol{\psi}, s) = \mathcal{R}\boldsymbol{f}(\boldsymbol{\psi}, s),$$

which completes the proof. $\qquad\square$

It is easy to see that if $\mathcal{Z}_1 \cup \mathcal{Z}_2$ is finite, then all singularities appearing on the left-hand side of formula (6.30) are removable. In other words, we have the following.

Remark 6.6. If $\{\boldsymbol{\zeta}^{(i)}\}_{i=1}^{M} = \mathcal{Z}_1 \cup \mathcal{Z}_2$, $M \geq m$, then by formula (6.32) and continuity of $\mathcal{R}\boldsymbol{f}$, we have

$$\lim_{\boldsymbol{\psi} \to \boldsymbol{\zeta}^{(i)}} \left[Q(\boldsymbol{\psi}) \frac{d}{ds} \mathcal{R}(\boldsymbol{Sf})(\boldsymbol{\psi}, s) \right] = \lim_{\boldsymbol{\psi} \to \boldsymbol{\zeta}^{(i)}} \left[Q(\boldsymbol{\psi}) Q(\boldsymbol{\psi})^{-1} \mathcal{R}\boldsymbol{f}(\boldsymbol{\psi}, s) \right]$$

$$= \mathcal{R}\boldsymbol{f}\left(\boldsymbol{\zeta}^{(i)}, s \right) < \infty.$$

Corollary 6.1. *If $\mathcal{Z}_1 \cup \mathcal{Z}_2$ is finite, then one can apply $\mathcal{R}^{-1}$ to both sides of equation (6.30) to recover $\boldsymbol{f}(\boldsymbol{x})$ at every $\boldsymbol{x} \in D_1$.*

It is clear that the set $\mathcal{Z}_1$ is finite. Let us study in more detail the set $\mathcal{Z}_2$, i.e. the solutions of the equation

$$\gamma(\psi) = -\sum_{i=1}^{m} \frac{c_i \gamma_i}{\psi \cdot \gamma_i} = 0. \tag{6.33}$$

Bringing the fractions in the above sum to a common denominator, we have

$$\gamma(\psi) = \frac{-\sum_{i=1}^{m} \left(\prod_{j\neq i} \psi \cdot \gamma_j \right) c_i \gamma_i}{\prod_{j=1}^{m} \psi \cdot \gamma_j}. \tag{6.34}$$

The numerator is a vector function, which we denote by $-\mathbf{P}(\psi)$.

Using the notation $\gamma_j = (a_j, b_j)$, $j = 1, \ldots, m$, we can rewrite $\mathbf{P}(\psi)$ in terms of the components of ψ as follows:

$$\mathbf{P}(\psi_1, \psi_2) = \sum_{i=1}^{m} \left[c_i \prod_{j\neq i} (\psi_1 a_j + \psi_2 b_j) \right] (a_i, b_i). \tag{6.35}$$

Note that each component P_i of $\mathbf{P} = (P_1, P_2)$ is a homogeneous polynomial in terms of $\psi = (\psi_1, \psi_2)$. It is not hard to see that on $\mathbb{S}^1 \setminus \mathcal{Z}_1$, the condition $\gamma(\psi) = 0$ reduces to $\mathbf{P}(\psi) = 0$, i.e. $P_1(\psi) = 0$ and $P_2(\psi) = 0$.

The classical Bézout's theorem from algebraic geometry implies that two affine algebraic plane curves C_1 and C_2 of degree d_1 and d_2, respectively, either have a common component (i.e. the polynomials defining them have a common factor) or intersect in at most $d_1 d_2$ points (e.g. see Fischer, 2001). Since $P_c(\psi_1, \psi_2) \doteq \psi_1^2 + \psi_2^2 - 1$ is irreducible, each homogeneous polynomial $P_i(\psi_1, \psi_2)$ either has finitely many zeros on $\mathbb{S}^1$ or is identically zero. In other words, the set $\mathcal{Z}_2$ is either finite or contains $\mathbb{S}^1 \setminus \mathcal{Z}_1$. Thus, we have the following.

Corollary 6.2. *The star transform is invertible if $\gamma(\psi) \neq 0$ for at least one $\psi \in \mathbb{S}^1 \setminus \mathcal{Z}_1$. In fact, the proof of Theorem 6.9 will show that the converse implication is also true, i.e. it is an "if and only if" statement.*

This corollary helps us to give a complete description of the invertible star configurations.

Definition 6.11. We call a star transform $\mathcal{S}$ *symmetric* if $m = 2k$ for some $k \in \mathbb{N}$ and (after possible re-indexing) $\gamma_i = -\gamma_{k+i}$ with $c_i = -c_{k+i}$ for all $i = 1, \ldots, k$.

As a side note, the sign convention in $c_i = -c_{k+i}$ is different from that of Chapter 5, which is due to the orientation that we are using in the definition of the star transform for vector fields.

Theorem 6.9. *The star transform $\mathcal{S}$ is invertible if and only if it is not symmetric.*

Proof. The argument follows closely the steps of the proof of Theorem 5.2, which we present here for completeness. If $\mathcal{S}$ is symmetric with $2k$ ray directions, then the star transform data contains less information than the data of standard Radon transform in k fixed directions (one can obtain the star transform from a sum of the corresponding pairs of the Radon transform). It is well known that we cannot recover a function from the Radon data of finitely many angles; therefore, $\mathcal{S}$ cannot be invertible.

Now, assume that for a fixed choice of $\gamma_1, \ldots, \gamma_m$, there is no inversion for the corresponding star transform. By Corollary 6.2, we have $\gamma(\psi) \equiv 0$. Hence, $\mathbf{P}(\psi_1, \psi_2) \equiv 0$.

Without loss of generality, assume that $a_1 \neq 0$ (otherwise $b_1 \neq 0$). If we take $\psi_1 = b_1$ and $\psi_2 = -a_1$, then formula (6.35) implies the following:

$$0 = a_1 \prod_{j \neq 1} (\psi_1 a_j + \psi_2 b_j) = a_1 (a_2 b_1 - a_1 b_2) \ldots (a_m b_1 - a_1 b_m).$$

Hence, for some index ℓ, we are required to have $a_\ell b_1 = a_1 b_\ell$ or, equivalently, $a_1/b_1 = a_\ell/b_\ell$. Given the assumption that γ_is are distinct unit vectors, we conclude that $\gamma_1 = -\gamma_\ell$. Applying this procedure with $\psi_1 = a_j$ and $\psi_2 = -a_j$ repeatedly for all j, we conclude that in order for the star transform to be noninvertible, its ray directions have to come in opposite pairs.

Now, we prove the relation $c_i = -c_{k+i}$ between the corresponding weights of each pair. Without loss of generality, let us assume that $m = 2k$, $\gamma_i = -\gamma_{k+i}$ for $i = 1, \ldots, k$ and no pair of vectors $\gamma_1, \ldots, \gamma_k$ are collinear. Then, we can rewrite formula (6.35) as

$$\mathbf{P}(\psi_1, \psi_2) = (-1)^k \prod_{j=1}^{k} (\psi_1 a_j + \psi_2 b_j) \sum_{i=1}^{k} \left[(c_i + c_{k+i}) \gamma_i \prod_{\substack{j=1 \\ j \neq i}}^{k} (\psi_1 a_j + \psi_2 b_j) \right]$$

$$= 0, \quad \forall \ \psi_1, \psi_2 \in \mathbb{R}.$$

Since $\prod_{j=1}^{k}(\psi_1 a_j + \psi_2 b_j) = (\boldsymbol{\psi} \cdot \boldsymbol{\gamma}_1) \ldots (\boldsymbol{\psi} \cdot \boldsymbol{\gamma}_k)$, we have

$$\boldsymbol{\gamma}(\boldsymbol{\psi}) = \frac{-\mathbf{P}(\psi_1, \psi_2)}{(\boldsymbol{\psi} \cdot \boldsymbol{\gamma}_1) \ldots (\boldsymbol{\psi} \cdot \boldsymbol{\gamma}_m)} \qquad (6.36)$$

$$= \frac{-1}{(\boldsymbol{\psi} \cdot \boldsymbol{\gamma}_1) \ldots (\boldsymbol{\psi} \cdot \boldsymbol{\gamma}_k)} \sum_{i=1}^{k} \left[(c_i + c_{k+i}) \boldsymbol{\gamma}_i \prod_{\substack{j=1 \\ j \neq i}}^{k} (\psi_1 a_j + \psi_2 b_j) \right]$$

for all $\boldsymbol{\psi} \in S^1$ outside of the finite set $\{\boldsymbol{\psi} : \boldsymbol{\psi} \cdot \boldsymbol{\gamma}_i = 0, \ i = 1, \ldots, m\}$.
Hence, in order for $\mathcal{S}$ to be noninvertible, we must have

$$\sum_{i=1}^{k} \left[(c_i + c_{k+i}) \boldsymbol{\gamma}_i \prod_{\substack{j=1 \\ j \neq i}}^{k} (\psi_1 a_j + \psi_2 b_j) \right] \equiv 0.$$

Following the argument from the first part of the proof, if we take $\psi_1 = b_1$ and $\psi_2 = -a_1$, then

$$0 = (c_1 + c_{k+1}) \prod_{j=2}^{k} (\psi_2 a_j + \psi_2 b_j)$$

$$= (c_1 + c_{k+1}) (a_2 b_1 - a_1 b_2) \ldots (a_k b_1 - a_1 b_k).$$

Since all vectors $\boldsymbol{\gamma}_1, \ldots, \boldsymbol{\gamma}_k$ are pairwise linearly independent, $a_j b_1 - a_1 b_j \neq 0$ for $j = 2, \ldots, k$. Hence, the last equation implies that $c_{k+1} = -c_1$.

Applying this procedure with $\psi_1 = a_j$ and $\psi_2 = -a_j$ repeatedly for all j, we conclude that in order for the star transform to be noninvertible, we must have $c_{k+i} = -c_i$ for all $i = 1, \ldots, k$. $\qquad \square$

Theorem 6.9 immediately implies the following.

Corollary 6.3. *Any vector-valued star transform with an odd number of rays is invertible.*

Remark 6.7. It is easy to note that similar to the case of V-line transforms, there exists a disc D_2 of finite radius $r_2 > 0$ with the property that the restriction of $\mathcal{S}f$ to $\overline{D_2}$ completely defines it in $\mathbb{R}^2$. In other words, to recover a vector field $\mathbf{f}$ in $\mathbb{R}^2$ with components in $C_c^2(D_1)$, one only needs to know its vector-valued star transform $\mathcal{S}f$ in D_2.

Remark 6.8. Similar to the case of the star transform of scalar functions studied in Chapter 5, in invertible configurations, the function $Q(\psi)$ (or, equivalently, the function $\gamma(\psi)$) contains information about the stability of inversion of the star transform on vector fields. The comprehensive analysis of these functions is still an open problem.

Remark 6.9. When $m = 2$ and $c_1 = -c_2 = 1$, the star transform of vector field $\mathbf{f}$ corresponds to the vector function $(\mathcal{L}\mathbf{f}, \mathcal{T}\mathbf{f})$. Hence, Theorem 6.8 provides another approach to recovering the full vector field $\mathbf{f}$ from its longitudinal and transverse V-line transforms. In the special case where $\gamma_1 = -\gamma_2$ (and only in that case), the matrix $Q(\psi)$ is undefined for any ψ and the corresponding transform is not invertible.

6.7 Additional Remarks

(1) There are some interesting similarities between V-line vector tomography and classical (straight line) vector tomography, despite the differences in the concepts and techniques of deriving the results.

- The kernel descriptions for the longitudinal and transverse transforms are identical in the V-line case presented here and the straight line case (see Derevtsov and Pickalov, 2011).
- In the article by Derevtsov and Pickalov (2011), the authors showed that the combination of LRT and TRT provides a unique reconstruction of a vector field in $\mathbb{R}^2$. As it was demonstrated in this chapter, the same result is achieved by combining the V-line versions of those transforms.
- Krishnan *et al.* (2019a) and Mishra (2020) used the combination of LRT and the first integral moment transform data to get the full vector field in $\mathbb{R}^2$. Our Theorem 6.6 achieves the same result in the case of V-line transforms.

(2) In dimensions $d \geq 3$, the longitudinal V-line transform can be defined in the same fashion as for $d = 2$, but the transverse V-line transform will require more details since there is a $(d-1)$-dimensional space of transverse directions. Once a proper choice for the transverse direction is made, techniques similar to the ones presented in this chapter can be used to study the injectivity and invertibility for both transforms in higher dimensions.

(3) A typical application of vector field tomography is the recovery of fluid flow from reciprocal measurements of certain signals through

the domain of the flow. Some examples of such measurements include changes in travel time of acoustic signals sent in opposite directions or changes in the optical path length (integral of the refraction index along the linear path) of a laser beam shined through the fluid (Norton, 1992). The latter model assumes that the photon beams are collimated and the changes in the (linear) optical path are expressed in the optical phase shift measured interferometrically. In this setup, the information carried by the reflected or scattered photons is lost. At the same time, if the fluid has small suspended particles with a different refractive index from the medium, they will cause some photons to change their direction due to reflection and scattering. It is conceivable that one can use a sensor array on the opposite (from the laser source) side of the flow region to measure the data carried by such reflected particles. In the case of single reflections, a potential model for such data can be based on the V-line transforms of the flow field.

Appendix A

Mathematical Tools

Here, we present some mathematical ideas and results that play an important role in the proofs of the theorems discussed in the book. At the same time, they are somewhat technical and not directly related to integral geometry, so placing them in the main part of the book could become an unnecessary distraction from its narrative.

A.1 Cone Differentiation and Integration

In this section, we derive a generalization of the fundamental theorem of calculus (FTC) to $\mathbb{R}^n$ from the perspective of partially ordered sets. We call this generalization *cone differentiation theorem* due to the concept of a positive cone in a partially ordered vector space (see the following).

To motivate our discussion, let us write the FTC in terms of the natural order in $\mathbb{R}$ and then build the necessary background for our generalization to $\mathbb{R}^n$.

Theorem A.10. *Let $\leq$ be the natural order on $\mathbb{R}$. If $f \in L^1(\mathbb{R})$ and*

$$F(x) = \int_{y \leq x} f(y)\, dy, \tag{A.1}$$

then $F' = f$ almost everywhere.

Remark A.10. Note that in this case, F is absolutely continuous (e.g. see Rudin, 1987).

A.1.1   *Partial order on* $\mathbb{R}^n$

Definition A.12. A *partially ordered vector space* V is a vector space over $\mathbb{R}$ together with a partial order $\leq$ such that:

(1) If $\mathbf{x} \leq \mathbf{y}$, then $\mathbf{x} + \mathbf{z} \leq \mathbf{y} + \mathbf{z}$ for all $\mathbf{z} \in V$.
(2) If $\mathbf{x} \geq 0$, then $c\mathbf{x} \geq 0$ for all $c \in \mathbb{R}^+$.

Definition A.13. The *positive cone* V^+ of vector space V is defined as

$$V^+ = \{\mathbf{x} \in V; \mathbf{x} \geq 0\}.$$

Definition A.12 implies that $\mathbf{x} \leq \mathbf{y} \Leftrightarrow 0 \leq \mathbf{x} - \mathbf{y}$, and hence, the partial order on V is completely determined by V^+.

For example, one can define a partial order in $\mathbb{R}^2$ as follows: $(x_1, x_2) \leq (y_1, y_2)$ if and only if $x_1 \leq y_1$ and $x_2 \leq y_2$. In this case, the positive cone will coincide with the first quadrant.

As another example, let us start with a choice of a positive cone in $\mathbb{R}^2$ and deduce the partial order from it. Consider $V^+ = \{(x_1, x_2) : \; x_1 \geq 0, x_2 \geq x_1\}$. Then, the corresponding partial order in $\mathbb{R}^2$ will be given by the following: $(x_1, x_2) \leq (y_1, y_2)$ if and only if $x_1 \leq y_1$ and $y_2 - x_2 \leq y_1 - x_1$.

In general, for $P \subset V$, there is a partial order on V such that $P = V^+$ if and only if

$$P \cap (-P) = \{0\},$$
$$P + P \subset P,$$
$$c \geq 0 \Rightarrow cP \subset P.$$

One can also identify a partial order structure with $V^- = -V^+$, which is called the *negative cone* of V.

These concepts play an important role in functional analysis and its applications. For a detailed treatment of the subject, we refer the reader to Fremlin (1974) and Holmes (1975). Here, we use the positive and negative cones to motivate and guide the generalization of certain classical results of analysis on the real line to higher dimensions.

Let us consider partial orders in $\mathbb{R}^n$ corresponding to positive (or negative) cones $C_\mathcal{V}$ generated by a set of fixed basis vectors $\mathcal{V} = \{\mathbf{v}_1, \ldots, \mathbf{v}_n\}$, i.e. $C_\mathcal{V} = \{\sum_{i=1}^n c_i \mathbf{v}_i; c_i \geq 0\}$.

In the case of $\mathbb{R}^2$, we use two linearly independent vectors $\mathbf{u}, \mathbf{v}$ as a generating set for the positive (or negative) cone. In this case, the boundary

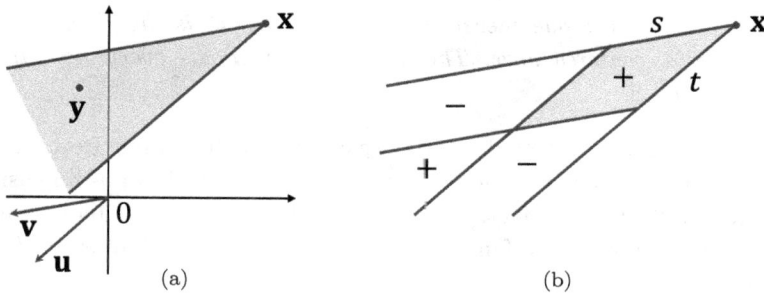

Fig. A.1 Negative cones and integration: (a) The negative cone at **x** with respect to a fixed partial order in $\mathbb{R}^2$ generated by vectors **u, v**; (b) combining the integrals of f over negative cones to get its integral over a parallelogram.

of the cone is a V-line. This fact is an important building block of our construction of the inversion formula for the VLT in Section 3.3.1.

In analogy with formula (A.1), for $f \in L^1(\mathbb{R}^n)$, we define F on $\mathbb{R}^n$ as

$$F(x) = \int_{y \leq x} f(y)\, d\mu, \tag{A.2}$$

where μ is the standard Lebesgue measure on $\mathbb{R}^n$ and $y \leq x$ represents the negative cone at **x** with respect to a fixed partial order on $\mathbb{R}^n$. In other words, the integral is taken over the region $\{y \in \mathbb{R}^n; y \leq x\}$ (see Figure A.1(a)).

A.1.2 *Two classical generalizations of FTC*

Now, a natural question is: In what sense of differentiation can one generalize the FTC to $\mathbb{R}^n$? While we construct a version of such a generalization using an order structure in $\mathbb{R}^n$ and its geometric properties, there are several classical results that are broadly considered as generalizations of the FTC. The derivations of our results rely on two such theorems, which we list in the following.

Theorem A.11 Lebesgue differentiation theorem, see Lebesgue (1910) and Rudin (1987). *Let* $f : \mathbb{R}^n \to \mathbb{R}$ *be integrable. Then, the following is true almost everywhere:*

$$f(x) = \lim_{r \to 0} \frac{1}{\mu(B_r(x))} \int_{B_r(x)} f d\mu,$$

where μ is the Lebesgue measure on $\mathbb{R}^n$ and $B_r(\boldsymbol{x})$ is the Euclidean ball with radius r centered at $\boldsymbol{x}$. The above equality holds everywhere if f is continuous on $\mathbb{R}^n$.

Note that this result holds even if we consider balls coming from another equivalent metric structure on $\mathbb{R}^n$, for example if $B_r(\mathbf{x})$ is an n-dimensional cube centered at $\mathbf{x}$. In fact, the family of balls described above can be replaced by a fairly large family of open sets that *"shrink to $\boldsymbol{x}$ nicely,"* as explained by Rudin (1987).

Before stating the second theorem, let us recall a definition from measure theory.

Definition A.14. Let μ be a positive measure defined on a σ-algebra $\mathfrak{M}$, and let ν be a signed measure on $\mathfrak{M}$. Then, ν is called *absolutely continuous with respect to μ*, denoted by

$$\nu \ll \mu,$$

if $\nu(E) = 0$ whenever $\mu(E) = 0$.

Theorem A.12 (Radon–Nikodym theorem, see Rudin (1987)). *If μ is the Lebesgue measure on $\mathbb{R}^n$ and ν is a signed measure on $\mathbb{R}^n$ such that $\nu \ll \mu$, then there is a unique integrable real-valued function f on $\mathbb{R}^n$ such that for every measurable set A,*

$$\nu(A) = \int_A f \, d\mu.$$

f is called the Radon–Nikodym derivative of ν with respect to μ.

Furthermore, if ν is a measure (nonnegative), then the function f will be a nonnegative function.

For more details about the "nicely shrinking sets," Lebesgue differentiation, and their relation to the Radon–Nikodym theorem, we refer the reader to Chapter 7 of Rudin (1987).

A.1.3 *Cone differentiation theorem*

In this section we derive another generalization of the FTC, which we use later to study the properties of the VLT and its generalizations.

We start with the 2D case. Assume f is an integrable function with respect to the Lebesgue measure on $\mathbb{R}^2$. Let $F(\mathbf{x})$ be defined as in formula

(A.2) using the partial order corresponding to the positive cone generated by some fixed vectors $\mathbf{u}, \mathbf{v}$.

Define $A_{t,s}(\mathbf{x})$ as the average of f over the parallelogram P centered at $\mathbf{x}$, with sides of length t, s and directions $\mathbf{u}, \mathbf{v}$. We can consider these parallelograms as a family of nicely shrinking neighborhoods described by Rudin (1987). Note that the area of the parallelogram generated by vectors $t\mathbf{u}, s\mathbf{v}$ is equal to $|\det(t\mathbf{u}, s\mathbf{v})| = ts\,|\det(\mathbf{u}, \mathbf{v})|$, and we have

$$A_{s,t}(\mathbf{x}) = \frac{1}{ts\,|\det(\mathbf{u}, \mathbf{v})|} \int_P f\, d\mu.$$

Using a simple geometric argument (see Figure A.1(b)) and the fact that $F(\mathbf{x})$ is the integral of f over the negative cone at $\mathbf{x}$, we get

$$A_{t,s}(\mathbf{x}) = \frac{1}{ts\,|\det(\mathbf{u}, \mathbf{v})|} \left[F\left(\mathbf{x} + \frac{t}{2}\mathbf{u} + \frac{s}{2}\mathbf{v}\right) - F\left(\mathbf{x} - \frac{t}{2}\mathbf{u} + \frac{s}{2}\mathbf{v}\right) \right.$$
$$\left. - F\left(\mathbf{x} + \frac{t}{2}\mathbf{u} - \frac{s}{2}\mathbf{v}\right) + F\left(\mathbf{x} - \frac{t}{2}\mathbf{u} - \frac{s}{2}\mathbf{v}\right) \right].$$

Likewise, for the n-dimensional case using a geometric argument and induction over n, we get the following averaging formula for f:

$$A_{t_1,\ldots,t_n}(\mathbf{x}) \tag{A.3}$$
$$= \frac{\sum_{\sigma\in\{-\frac{1}{2},\frac{1}{2}\}^n} \mathrm{sgn}(\sigma_1\ldots\sigma_n)\, F(\mathbf{x} + \sigma_1 t_1 \mathbf{v}_1 + \cdots + \sigma_n t_n \mathbf{v}_n)}{t_1\ldots t_n\,|\det(\mathbf{v}_1,\ldots,\mathbf{v}_n)|}.$$

In the special case where $t_1 = \cdots = t_n = t$, this quantity corresponds to the average of f over P_t, the parallelepiped with sides of length t centered at $\mathbf{x}$, which we denote it by $A_t(\mathbf{x})$, i.e.

$$A_t(\mathbf{x}) = \frac{1}{\mu(P_t)} \int_{P_t} f\, d\mu = \frac{1}{t^n\,|\det(\mathbf{v}_1,\ldots,\mathbf{v}_n)|} \int_{P_t} f\, d\mu. \tag{A.4}$$

Averaging over such infinitesimal symmetric neighborhoods of $\mathbf{x}$ and applying Theorem A.11, we obtain the following result:

Theorem A.13. *Let $\leq$ be an order structure in $\mathbb{R}^n$ corresponding to the positive cone generated by vectors $v_1,\ldots,v_n$. Let $f \in L^1(\mathbb{R}^n)$ and $F(x)$ be defined as in formula (A.2). Then, for almost every x, we have*

$$f(x) = \lim_{t\to 0} A_t(x), \tag{A.5}$$

where $A_t(x)$ is defined in formula (A.4) and can be computed using $t_1 = \cdots = t_n = t$ in formula (A.3).

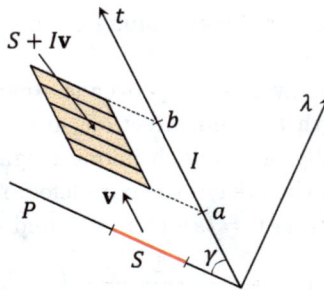

Fig. A.2   A sketch of the sections of $S + Iv$.

Note that this method of recovering f from F is of practical significance because it is both efficient and simple.

Before introducing the main theorem of this section, we prove one more technical result. In essence, it is a special case of Fubini's theorem, but it is used multiple times throughout the book, so we prove it here and give it a geometrically descriptive name.

Let P be a hyperplane in $\mathbb{R}^n$ and $S \subset P$ be a measurable set. For a vector $\mathbf{v} \in \mathbb{R}^n$ transversal to P, we denote $S + t\mathbf{v} = \{\mathbf{s} + t\mathbf{v} : \mathbf{s} \in S\}$, and for $I = [a, b] \subset \mathbb{R}$, we let $S + I\mathbf{v} = \{S + t\mathbf{v} : t \in [a, b]\}$. Geometrically, $S + t\mathbf{v}$ is an $n-1$-dimensional section of the set $S + I\mathbf{v}$ (see Figure A.2).

Lemma A.1 (Moving sections lemma). *Let $f \in L^1(\mathbb{R}^n)$ and $P, S, \mathbf{v}$ be defined as above. Then,*

$$\int_{S+I\mathbf{v}} f \, d\mu = \sin\gamma \int_a^b \int_{S+t\mathbf{v}} f \, d\mu_P \, dt,$$

where μ_P is induced from the natural measure on P and γ is the angle between $\mathbf{v}$ and S.

Proof. Let λ be the coordinate measured along the axis normal to P. Then, the Lebesgue measure on $\mathbb{R}^n$ is the product of the Lebesgue measure on the λ-axis and the natural Lebesgue measure on P. Now, let t be the coordinate measured along the axis in $\mathbf{v}$-direction. Then, we have $d\lambda = \sin\gamma \, dt$.

By Fubini's theorem,

$$\int_{S+I\mathbf{v}} f \, d\mu = \int_{a\sin\gamma}^{b\sin\gamma} \int_{S+\lambda\sin\gamma\,\mathbf{v}} f \, d\mu_P \, d\lambda = \int_a^b \int_{S+t\mathbf{v}} f \, d\mu_P \sin\gamma \, dt.$$

$\square$

In particular, as a consequence of this lemma in $\mathbb{R}^2$, we have the following.

Corollary A.4. *Let S be a line segment in $\mathbb{R}^2$ and v be a vector transversal to S. If the integral of f along the section $S + tv$ is denoted by*

$$h(t) = \int_{S+tv} f \, dl,$$

then the integral of f over the parallelogram $S + Iv$ can be computed as

$$\int_{S+Iv} f \, d\mu = \sin \gamma \int_a^b h(t) \, dt.$$

Let us now state the main result of this section.

Theorem A.14. (*cone differentiation theorem*) *Let $\leq$ be an order structure in $\mathbb{R}^n$ corresponding to the positive cone generated by unit vectors $v_1, \ldots, v_n$. Assume $f \in C_c(\mathbb{R}^n)$ and $F(x) = \int_{y \leq x} f(y) d\mu$. Then, we have*

$$f(x) = \frac{1}{|\det(v_1, \ldots, v_n)|} D_{v_1} \ldots D_{v_n} F(x), \qquad (A.6)$$

where D_{v_j} is the directional derivative in the direction of v_j.

Proof. We provide two different proofs here. The first one is a geometrically intuitive argument applicable to functions in $\mathbb{R}^2$, which is the most relevant case for imaging applications described in the book. The second proof is more general and covers arbitrary dimensions.

Proof 1: Assume $f \in C_c(\mathbb{R}^2)$, and let $\mathbf{u}$ and $\mathbf{v}$ be the unit vectors generating the positive cone in $\mathbb{R}^2$. For $t \in \mathbb{R}$ and $\mathbf{x} \in \mathbb{R}^2$, we define

$$q_{\mathbf{x}}(t) = \int_{-\infty}^0 f(\mathbf{x} + t\mathbf{u} + s\mathbf{v}) \, ds.$$

In geometric terms, $q_{\mathbf{x}}(t)$ is the integral of f over the ray propagating from the point $\mathbf{x} + t\mathbf{u}$ in the direction opposite to $\mathbf{v}$. Since $f \in C_c(\mathbb{R}^2)$, one can conclude that $q_{\mathbf{x}}$ is also continuous, e.g. by using the dominated convergence theorem with the dominant L^1 function $\chi_{[-d,d]} \max |f|$, where $d = \operatorname{diam} \operatorname{supp} f$ and χ_M is the characteristic function of the set M.

Recall that F is the integral of f over the negative cone. Hence, by the moving sections lemma, we have

$$F(\mathbf{x}) = \sin(2\beta) \int_{-\infty}^{0} q_{\mathbf{x}}(t)\, dt,$$

where 2β is the angle between $\mathbf{u}$ and $\mathbf{v}$. Now, using the Lebesgue differentiation theorem and continuity of $q_{\mathbf{x}}$, we obtain

$$D_{\mathbf{u}}F(\mathbf{x}) = \lim_{h \to 0} \frac{F(\mathbf{x} + h\mathbf{u}) - F(\mathbf{x})}{h} = \lim_{h \to 0} \frac{\sin(2\beta)}{h} \int_{0}^{h} q_{\mathbf{x}}(t)\, dt$$

$$= \sin(2\beta)\, q_{\mathbf{x}}(0).$$

Using the fact that $\sin(2\beta) = |\det(\mathbf{u}, \mathbf{v})|$ and applying the Lebesgue differentiation theorem once again, we get

$$\frac{1}{|\det(\mathbf{u}, \mathbf{v})|} D_{\mathbf{v}}D_{\mathbf{u}}F(x) = \lim_{h \to 0} \frac{q_{\mathbf{x}+h\mathbf{v}}(0) - q_{\mathbf{x}}(0)}{h}$$

$$= \lim_{h \to 0} \frac{1}{h} \int_{0}^{h} f(\mathbf{x} + t\mathbf{v})\, dt = f(\mathbf{x}).$$

Proof 2: Consider the linear transformation $\phi : \mathbb{R}^n \to \mathbb{R}^n$ defined by $\phi(\mathbf{e}_i) = \mathbf{v}_i$, $i = 1, \ldots, n$, where $\{\mathbf{e}_i\}_{i=1}^{n}$ is the standard basis for $\mathbb{R}^n$. In the domain of ϕ, consider the partial order corresponding to the positive cone generated by $\{\mathbf{e}_i\}_{i=1}^{n}$. In the range of ϕ, consider the partial order corresponding to the positive cone generated by $\{\mathbf{v}_i\}_{i=1}^{n}$.

Denote by $C(\hat{\mathbf{x}}) = \{\hat{\mathbf{y}} \in \mathbb{R}^n \,|\, \hat{\mathbf{y}} \le \hat{\mathbf{x}}\}$ the negative cone at $\hat{\mathbf{x}}$ in the domain of ϕ. Then, $\phi(C(\hat{\mathbf{x}})) = \{\mathbf{y} \in \mathbb{R}^n,\ \mathbf{y} \le \mathbf{x}\}$ is the negative cone at $\mathbf{x} = \phi(\hat{\mathbf{x}})$ in the range of ϕ.

Using the change of variables formula for n-dimensional integrals, we obtain

$$F(\mathbf{x}) = \int_{\mathbf{y} \le \mathbf{x}} f(\mathbf{y})\, d\mathbf{y} = \int_{\phi(C(\hat{\mathbf{x}}))} f(\mathbf{y})\, d\mathbf{y} = \int_{C(\hat{\mathbf{x}})} f(\phi(\hat{\mathbf{y}}))\, |\det(\phi)|\, d\hat{\mathbf{y}}.$$

Now, for $\mathbf{x}_0 = \phi(\hat{\mathbf{x}}_0)$, we have

$$D_{\mathbf{v}_1} \ldots D_{\mathbf{v}_n} F(\mathbf{x}_0) = D_{\mathbf{v}_1} \ldots D_{\mathbf{v}_n} \left[\int_{\mathbf{y} \le \mathbf{x}} f(\mathbf{y})\, d\mathbf{y} \right](\mathbf{x}_0)$$

$$= D_{\mathbf{e}_1} \ldots D_{\mathbf{e}_n} \left[\int_{C(\hat{\mathbf{x}})} f(\phi(\hat{\mathbf{y}}))\, |\det(\phi)|\, d\hat{\mathbf{y}} \right](\hat{\mathbf{x}}_0)$$

$$= |\det(\phi)| \frac{\partial^n}{\partial \hat{x}_1 \ldots \partial \hat{x}_n} \bigg|_{\hat{\mathbf{x}}_0} \int_{-\infty}^{\hat{x}_n} \ldots \int_{-\infty}^{\hat{x}_1} f(\phi(\hat{\mathbf{y}}))\, d\hat{y}_1 \ldots d\hat{y}_n$$

$$= |\det(\phi)|\, f(\phi(\hat{\mathbf{x}}_0)) = |\det(\mathbf{v}_1, \ldots, \mathbf{v}_n)|\, f(\mathbf{x}_0),$$

$$\text{(A.7)}$$

where we used Fubini's theorem and the FTC (n times) correspondingly in the third and fourth lines. □

Corollary A.5. *In $\mathbb{R}^2$, the cone differentiation theorem can be written as*

$$f(\boldsymbol{x}) = \frac{1}{|\det(\boldsymbol{u}, \boldsymbol{v})|} D_u D_v F(\boldsymbol{x}). \qquad (A.8)$$

Remark A.11. Note that Theorem A.14 does not imply that $F \in C^n(\mathbb{R}^n)$. For example, $D_v^2 F$ may not exist in $\mathbb{R}^2$.

A.2 Absolute Continuity

In the case of the finite measure ν, Definition A.14 of absolute continuity is equivalent to the following (e.g. see Rudin, 1987, Theorem 6.11)):

Definition A.15. A measure ν is absolutely continuous with respect to Lebesgue measure μ if for any $\epsilon > 0$, there exists a $\delta > 0$ such that $\mu(E) < \delta$ implies $\nu(E) < \epsilon$.

Definition A.16. A cumulative distribution function F is called absolutely continuous if for any $\epsilon > 0$, there exists a $\delta > 0$ such that $\sum \mu(P_i) < \delta$ implies $\sum \nu_F(P_i) < \epsilon$ for any finite collection of disjoint parallelograms $\{P_i\}$. Here, the induced measure ν_F on parallelograms is defined using the cumulative distribution function F.

We want to prove the following statement (e.g. see Royden and Fitzpatrick, 2010, Section 20.3, Proposition 26):

Proposition A.1. *F is absolutely continuous $\iff$ ν_F induced by F is absolutely continuous with respect to the Lebesgue measure μ.*

Proof. ($\Rightarrow$) Let $\epsilon > 0$ be given. Then, by the assumption of absolute continuity of F, we can choose $\delta > 0$ such that $\sum \mu(P_i) < \delta$ implies $\sum \nu_F(P_i) < \epsilon/2$ for any finite collection of disjoint parallelograms $\{P_i\}$. Now, let $E \subset \mathbb{R}^n$ with $\mu(E) < \delta/2$. Then, there exists a countable disjoint collection of parallelograms $\{P_i\}$ such that $E \subset \bigcup_{i=1}^{\infty} P_i$ and $\mu\left(\bigcup_{i=1}^{\infty} P_i\right) = \sum_{i=1}^{\infty} \mu(P_i) < \delta$. For the measure ν_F, we have

$$\nu_F\left(\bigcup_{i=1}^{n} P_i\right) = \sum_{i=1}^{n} \nu_F(P_i) < \epsilon/2,$$

and hence,

$$\nu_F\left(\bigcup_{i=1}^{\infty} P_i\right) = \lim_{n\to\infty} \nu_F\left(\bigcup_{i=1}^{n} P_i\right) = \lim_{n\to\infty} \sum_{i=1}^{n} \nu_F(P_i) \le \epsilon/2.$$

Finally, we have

$$\nu_F(E) \le \nu_F\left(\bigcup_{i=1}^{\infty} P_i\right) < \epsilon.$$

($\Leftarrow$) Given $\epsilon > 0$, pick $\delta > 0$ using the definition of absolute continuity for ν_F. Let $\{P_i\}$ be any disjoint collection of parallelograms with $\sum \mu(P_i) < \delta$, hence $\mu(\cup P_i) < \delta$. Then, by the hypothesis, $\sum \nu_F(P_i) = \nu_F(\cup P_i) < \epsilon$. $\square$

A.3   VLT as a Map on $L^1(\mathbb{R}^2)$

A ray have measure zero as a subset of $\mathbb{R}^2$, hence changing the values of a function $f \in L^1(\mathbb{R}^2)$ along a ray will produce an equivalent function in L^1 sense. Here, we show that the ray transform respects this equivalence relation, i.e. the ray transform maps two equivalent functions to the same equivalence class. The same argument will work for VLT, as it consists of a sum of two ray transforms along fixed directions. Without loss of generality, we prove the statement for the case of vertical rays.

Assume $f, g \in L^1(\mathbb{R}^2)$ and $f = g$ almost everywhere. For an arbitrary $c \in \mathbb{R}$, consider the formal notations

$$\phi_c(x) = \int_c^{\infty} f(x,y)\, dy,$$

$$\psi_c(x) = \int_c^{\infty} g(x,y)\, dy.$$

By Fubini's theorem for any real numbers $a < b$, we have

$$\int_{[a,b]\times[c,\infty]} f\, d\mu = \int_a^b \int_c^{\infty} f(x,y)\, dy\, dx = \int_a^b \phi_c(x)\, dx,$$

$$\int_{[a,b]\times[c,\infty]} g\, d\mu = \int_a^b \int_c^{\infty} g(x,y)\, dy\, dx = \int_a^b \psi_c(x)\, dx,$$

where $\phi_c(x)$ and $\psi_c(x)$ are integrable functions of x and

$$\int_a^b \phi_c(x)\,dx = \int_a^b \psi_c(x)\,dx.$$

By the Lebesgue differentiation theorem, we get $\phi_c(x) = \psi_c(x)$ for almost every x. Moreover, this statement is true for all $c \in \mathbb{R}$.

Now, let us recall some facts about product measure. Let (X_1, Σ_1, μ_1) and (X_2, Σ_2, μ_2) be σ-finite measure spaces. Then, the product measure on the product measurable space satisfies the following:

$$(\mu_1 \times \mu_2)(E) = \int_{X_2} \mu_1(E^y)\,d\mu_2(y),$$

where $E^y = \{x \in X_1 \,|\, (x,y) \in E\}$.

Hence, if for some $E \subseteq \mathbb{R}^2$, we have $(\mu_1 \times \mu_2)(E) \neq 0$, then $\mu_1(E^y) \neq 0$ for some y. Now, if we let $E = \{(x,c) \in \mathbb{R}^2 \,|\, \phi_c(x) \neq \psi_c(x)\}$, this will imply that E is of measure zero.

In other words, we just showed that if $f, g \in L^1(\mathbb{R}^2)$ and $f = g$ almost everywhere in $\mathbb{R}^2$, then the values of the ray transform of these two functions coincide for almost every (vertical) ray in the plane.

Bibliography

Abhishek, A. (2020). Support theorems for the transverse ray transform of tensor fields of rank m, *Journal of Mathematical Analysis and Applications* **485**(2), 123828.

Abhishek, A. and Mishra, R. K. (2019). Support theorems and an injectivity result for integral moments of a symmetric m-tensor field, *Journal of Fourier Analysis and Applications* **25**(4), 1487–1512.

Abramowitz, M. and Stegun, I. A. (1964). *Handbook of Mathematical Functions, with Formulas, Graphs, and Mathematical Tables* (Dover, New York).

Achmad, B. and Hussein, E. M. (2004). An X-ray Compton scatter method for density measurement at a point within an object, *Applied Radiation and Isotopes* **60**(6), 805–814.

Adejumo, O., Balogun, F., and Egbedokun, G. (2011). Developing a Compton scattering tomography system for soil studies: theory, *Journal of Sustainable Development and Environmental Protection* **1**(3), 73–81.

Agranovsky, M., Finch, D., and Kuchment, P. (2009). Range conditions for a spherical mean transform, *Inverse Problems & Imaging* **3**(3), 373.

Agranovsky, M., Kuchment, P., and Quinto, E. T. (2007). Range descriptions for the spherical mean Radon transform, *Journal of Functional Analysis* **248**(2), 344–386.

Agranovsky, M. and Nguyen, L. V. (2010). Range conditions for a spherical mean transform and global extendibility of solutions of the Darboux equation, *Journal d'Analyse Mathématique* **112**(1), 351–367.

Aguilar, V. and Kuchment, P. (1995). Range conditions for the multidimensional exponential X-ray transform, *Inverse Problems* **11**(5), 977.

Aguilar, V., Ehrenpreis, L., and Kuchment, P. (1996). Range conditions for the exponential Radon transform, *Journal d'Analyse Mathématique* **68**(1), 1–13.

Aldawood, S., Thirolf, P., Miani, A., Böhmer, M., Dedes, G., Gernhäuser, R., Lang, C., Liprandi, S., Maier, L., Marinšek, T., *et al.* (2017). Development of a Compton camera for prompt-gamma medical imaging, *Radiation Physics and Chemistry* **140**, 190–197.

Allmaras, M., Darrow, D., Hristova, Y., Kanschat, G., and Kuchment, P. (2013). Detecting small low emission radiating sources, *Inverse Problems & Imaging* **7**(1), 47–79.

Alpuche Aviles, J. E., Pistorius, S., Gordon, R., and Elbakri, I. A. (2011). A novel hybrid reconstruction algorithm for first generation incoherent scatter CT (ISCT) of large objects with potential medical imaging applications, *Journal of X-Ray Science and Technology* **19**(1), 35–56.

Alvarez, R. E. and Macovski, A. (1976). Energy-selective reconstructions in X-ray computerised tomography, *Physics in Medicine & Biology* **21**(5), 733–744.

Ambartsoumian, G. (2012). Inversion of the V-line Radon transform in a disc and its applications in imaging, *Computers & Mathematics with Applications* **64**(3), 260–265.

Ambartsoumian, G. and Krishnan, V. P. (2015). Inversion of a class of circular and elliptical Radon transforms, in *Contemporary Mathematics*, Vol. 653, pp. 1–12.

Ambartsoumian, G. and Kuchment, P. (2006). A range description for the planar circular Radon transform, *SIAM Journal on Mathematical Analysis* **38**(2), 681–692.

Ambartsoumian, G. and Latifi, M. J. (2019). The V-line transform with some generalizations and cone differentiation, *Inverse Problems* **35**(3), 034003.

Ambartsoumian, G. and Latifi, M. J. (2021). Inversion and symmetries of the star transform, *The Journal of Geometric Analysis* **31**, 11270–11291.

Ambartsoumian, G. and Moon, S. (2013). A series formula for inversion of the V-line Radon transform in a disc, *Computers & Mathematics with Applications* **66**(9), 1567–1572.

Ambartsoumian, G. and Roy, S. (2016). Numerical inversion of a broken ray transform arising in single scattering optical tomography, *IEEE Transactions on Computational Imaging* **2**(2), 166–173.

Ambartsoumian, G., Gouia-Zarrad, R., and Lewis, M. A. (2010). Inversion of the circular Radon transform on an annulus, *Inverse Problems* **26**(10), 105015.

Ambartsoumian, G., Latifi, M. J., and Mishra, R. K. (2020). Generalized V-line transforms in 2D vector tomography, *Inverse Problems* **36**(10), 104002.

Anastasio, M. A., Pan, X., and Clarkson, E. (2001). Comments on the filtered backprojection algorithm, range conditions, and the pseudoinverse solution, *IEEE Transactions on Medical Imaging* **20**(6), 539–542.

Andreyev, A., Sitek, A., and Celler, A. (2009). Stochastic image reconstruction method for Compton camera, in *2009 IEEE Nuclear Science Symposium Conference Record (NSS/MIC)* (IEEE), 2985–2988.

Anghaie, S., Humphries, L. L., and Diaz, N. J. (1990). Material characterization and flaw detection, sizing, and location by the differential gamma scattering spectroscopy technique. Part I: Development of theoretical basis, *Nuclear Technology* **91**(3), 361–375.

Antich, P., Parkey, R., Slavin, N., Tsyganov, E., and Zinchenko, A. (2000). Compact Compton camera design: parameters and imaging algorithms, in *2000 IEEE Nuclear Science Symposium. Conference Record (Cat. No. 00CH37149)*, Vol. 3 (IEEE), 20–60.

Arbuzov, E., Bukhgeim, A., and Kazantsev, S. (1998). Two-dimensional tomography problems and the theory of A-analytic functions, *Siberian Advances in Mathematics* **8**(4), 1–20.

Arendtsz, N. V. and Hussein, E. M. (1993). Electron density tomography with compton-scattered radiation, in *Mathematical Methods in Medical Imaging II*, Vol. 2035 (International Society for Optics and Photonics), pp. 230–241.

Arendtsz, N. V. and Hussein, E. M. (1995a). Energy-spectral Compton scatter imaging. I. Theory and mathematics, *IEEE Transactions on Nuclear Science* **42**(6), 2155–2165.

Arendtsz, N. V. and Hussein, E. M. (1995b). Energy-spectral Compton scatter imaging. II. Experiments, *IEEE Transactions on Nuclear Science* **42**(6), 2166–2172.

Arridge, S. R. and Schotland, J. C. (2009). Optical tomography: forward and inverse problems, *Inverse Problems* **25**(12), 123010.

Arsenault, P. J. and Hussein, E. M. (2006). Image reconstruction from the compton scattering of X-ray fan beams in thick/dense objects, *IEEE Transactions on Nuclear Science* **53**(3), 1622–1633.

Athreya, K. B. and Lahiri, S. N. (2006). *Measure Theory and Probability Theory* (Springer-Verlag, New York).

Baines, W. (2021). Range description of a conical Radon transform, *arXiv preprint arXiv:2110.11212* .

Bal, G. (2009). Inverse transport theory and applications, *Inverse Problems* **25**(5), 053001.

Balogun, F. and Cruvinel, P. (2003). Compton scattering tomography in soil compaction study, *Nuclear Instruments and Methods in Physics Research Section A: Accelerators, Spectrometers, Detectors and Associated Equipment* **505**(1–2), 502–507.

Balogun, F. and Spyrou, N. (1993). Compton scattering tomography in the study of a dense inclusion in a lighter matrix, *Nuclear Instruments and Methods in Physics Research Section B: Beam Interactions with Materials and Atoms* **83**(4), 533–538.

Bandstra, M. S., Bellm, E. C., Boggs, S. E., Perez-Becker, D., Zoglauer, A., Chang, H.-K., Chiu, J.-L., Liang, J.-S., Chang, Y.-H., *et al.* (2011). Detection and imaging of the Crab Nebula with the nuclear Compton telescope, *The Astrophysical Journal* **738**(1), 8.

Barrett, H. and Myers, K. (2003). *Foundations of Image Science*, Wiley Series in Pure and Applied Optics (Wiley, Hoboken, NJ, USA).

Barrett, H. H., Gallas, B., Clarkson, E., and Clough, A. (1999). Scattered radiation in nuclear medicine: A case study on the Boltzmann transport equation, in *Computational Radiology and Imaging, The IMA Volumes in Mathematics and its Applications*, Vol. 110 (Springer), 71–100.

Basko, R., Zeng, G. L., and Gullberg, G. T. (1997a). Analytical reconstruction formula for one-dimensional Compton camera, *IEEE Transactions on Nuclear Science* **44**(3), 1342–1346.

Basko, R., Zeng, G. L., and Gullberg, G. T. (1997b). Fully three dimensional image reconstruction from "V"-projections acquired by Compton camera with three vertex electronic collimation, in *1997 IEEE Nuclear Science Symposium Conference Record*, Vol. 2, 1077–1081.

Basko, R., Zeng, G. L., and Gullberg, G. T. (1998). Application of spherical harmonics to image reconstruction for the Compton camera, *Physics in Medicine & Biology* **43**(4), 887.

Battista, J. and Bronskill, M. (1978). Compton-scatter tissue densitometry: calculation of single and multiple scatter photon fluences, *Physics in Medicine & Biology* **23**(1), 1.

Battista, J. and Bronskill, M. (1981). Compton scatter imaging of transverse sections: an overall appraisal and evaluation for radiotherapy planning, *Physics in Medicine & Biology* **26**(1), 81.

Battista, J., Santon, L., and Bronskill, M. (1977). Compton scatter imaging of transverse sections: corrections for multiple scatter and attenuations, *Physics in Medicine & Biology* **22**(2), 229.

Berker, Y. and Schulz, V. (2014). Scattered PET data for attenuation-map reconstruction in PET/MRI: Fundamentals, in *2014 IEEE Nuclear Science Symposium and Medical Imaging Conference (NSS/MIC)*, pp. 1–6.

Berodias, M. and Peix, M. (1988). Nondestructive measurement of density and effective atomic number by photon scattering, *Materials Evaluation* **46**(9), 1209–1213.

Bodette, D. and Jacobs, A. (1984). Tomographic two-phase flow distribution measurement using Gamma-ray scattering, in *New Technology in Nuclear Power Plant Instrumentation and Control* (Instrument Society of America), 23–34.

Boggs, S. and Jean, P. (2000). Event reconstruction in high resolution Compton telescopes, *Astronomy and Astrophysics Supplement Series* **145**(2), 311–321.

Boggs, S., Bandstra, M., Bowen, J., Coburn, W., Lin, R., Wunderer, C., Zoglauer, A., Amman, M., Luke, P., Jean, P., *et al.* (2005). Performance of the nuclear Compton telescope, *Experimental Astronomy* **20**(1-3), 387–394.

Bolozdynya, A. and Morgunov, V. (1998). Multilayer electroluminescence camera: concept and Monte Carlo study, *IEEE Transactions on Nuclear Science* **45**(3), 1646–1655.

Bolozdynya, A., Ordonez, C., and Chang, W. (1997b). A concept of cylindrical Compton camera for SPECT, in *1997 IEEE Nuclear Science Symposium Conference Record*, Vol. 2 (IEEE), 1047–1051.

Bolozdynya, A., Egorov, V., Koutchenkov, A., Safronov, G., Smirnov, G., Medved, S., and Morgunov, V. (1997a). High pressure xenon electronically collimated camera for low energy gamma ray imaging, *IEEE Transactions on Nuclear Science* **44**(6), 2408–2414.

Boman, J. and Strömberg, J.-O. (2004). Novikov's inversion formula for the attenuated Radon transform — a new approach, *The Journal of Geometric Analysis* **14**(2), 185–198.

Borcea, L., Papanicolaou, G., Tsogka, C., and Berryman, J. (2002). Imaging and time reversal in random media, *Inverse Problems* **18**(5), 1247.

Borg, L., Frikel, J., Jørgensen, J. S., and Quinto, E. T. (2018). Analyzing reconstruction artifacts from arbitrary incomplete X-ray CT data, *SIAM Journal on Imaging Sciences* **11**(4), 2786–2814.

Born, M. and Wolf, E. (2013). *Principles of Optics: Electromagnetic Theory of Propagation, Interference and Diffraction of Light* (Cambridge University Press, Cambridge, UK).

Brateman, L., Jacobs, A. M., and Fitzgerald, L. T. (1984). Compton scatter axial tomography with X-rays: SCAT-CAT, *Physics in Medicine & Biology* **29**(11), 1353.

Brechner, R. R. and Singh, M. (1990). Iterative reconstruction of electronically collimated SPECT images, *IEEE Transactions on Nuclear Science* **37**(3), 1328–1332.

Brunetti, A., Cesareo, R., Golosio, B., Luciano, P., and Ruggero, A. (2002). Cork quality estimation by using Compton tomography, *Nuclear Instruments and Methods in Physics Research Section B: Beam Interactions with Materials and Atoms* **196**(1-2), 161–168.

Busono, P. and Hussein, E. M. A. (1999). Algorithms for density and composition-discrimination imaging for fourth-generation CT systems, *Physics in Medicine & Biology* **44**(6), 1455–1477.

Cakoni, F. and Colton, D. (2006). *Qualitative Methods in Inverse Scattering Theory: An Introduction* (Springer Berlin, Heidelberg).

Cebeiro, J. and Morvidone, M. (2013). SVD inversion for the bi-dimensional conical Radon transform, *Journal of Physics: Conference Series* **477**(1), 012022.

Cebeiro, J., Morvidone, M., and Nguyen, M. (2016). Back-projection inversion of a conical Radon transform, *Inverse Problems in Science and Engineering* **24**(2), 328–352.

Cebeiro, J., Nguyen, M. K., Morvidone, M. A., and Noumowé, A. (2017). New "improved" Compton scatter tomography modality for investigative imaging of one-sided large objects, *Inverse Problems in Science and Engineering* **25**(11), 1676–1696.

Cebeiro, J., Tarpau, C., Morvidone, M. A., Rubio, D., and Nguyen, M. K. (2021). On a three-dimensional Compton scattering tomography system with fixed source, *Inverse Problems* **37**(5), 054001.

Cesareo, R., Borlino, C. C., Brunetti, A., Golosio, B., and Castellano, A. (2002). A simple scanner for Compton tomography, *Nuclear Instruments and Methods in Physics Research Section A: Accelerators, Spectrometers, Detectors and Associated Equipment* **487**(1-2), 188–192.

Chechkin, G. A. and Goritsky, A. Y. (2009). S. N. Kruzhkov's lectures on first-order quasilinear PDEs, in E. Emmrich and P. Wittbold (eds.), *Analytical and Numerical Aspects of Partial Differential Equations: Notes of a Lecture Series* (De Gruyter), 1–68.

Chelikani, S., Gore, J., and Zubal, G. (2004). Optimizing Compton camera geometries, *Physics in Medicine & Biology* **49**(8), 1387.

Cheney, M. and Borden, B. (2009). *Fundamentals of Radar Imaging* (SIAM).

Cheney, W. (2001). *Analysis for Applied Mathematics*, Graduate Texts in Mathematics (Springer-Verlag New York).

Clarke, R. L. (1965). A gamma ray scanner for diagnostic radiography, Tech. Rep. 2270, Atomic Energy of Canada Ltd, Chalk River, Ontario.

Clarke, R. L. and Van Dyk, G. G. (1969). Compton-scattered gamma rays in diagnostic radiography, in *Proceedings of a Symposium on Medical Radioisotope Scintigraphy*, Vol. I (Vienna: IAEA), 247–260.

Clarke, R. L. and Van Dyk, G. G. (1973). A new method for measurement of bone mineral content using both transmitted and scattered beams of gamma-rays, *Physics in Medicine & Biology* **18**(4), 532.

Clarke, R. L., Milne, E. N. C., and Van Dyk, G. G. (1976). The use of Compton scattered gamma rays for tomography. *Investigative Radiology* **11**(3), 225–235.

Clarkson, E. (1999). Projections onto the range of the exponential Radon transform and reconstruction algorithms, *Inverse Problems* **15**(2), 563.

Clinthorne, N. H., Ng, C.-Y., Hua, C.-H., Gormley, J., Leblanc, J., Wilderman, S. J., and Rogers, W. L. (1996). Theoretical performance comparison of a Compton-scatter aperture and parallel-hole collimator, in *1996 IEEE Nuclear Science Symposium. Conference Record*, Vol. 2 (IEEE), 788–792.

Colton, D. L. and Kress, R. (2013a). *Inverse Acoustic and Electromagnetic Scattering Theory*, Vol. 93 (Springer, New York, NY).

Colton, D. L. and Kress, R. (2013b). *Integral Equation Methods in Scattering Theory* (SIAM).

Compton, A. H. (1923). A quantum theory of the scattering of X-rays by light elements, *Physical Review* **21**(5), 483.

Conflitti, A. (2006). Zeros of real symmetric polynomials, *Applied Mathematics E-Notes* **6**, 219–224.

Cong, W. and Wang, G. (2011). X-ray scattering tomography for biological applications, *Journal of X-Ray Science and Technology* **19**, 219–227.

Conka-Nurdan, T., Nurdan, K., Constantinescu, F., Freisleben, B., Pavel, N., and Walenta, A. (2002). Impact of the detector parameters on a Compton camera, *IEEE Transactions on Nuclear Science* **49**(3), 817–821.

Cooper, M., Mijnarends, P., Shiotani, N., Sakai, N., and Bansil, A. (2004). *X-Ray Compton Scattering*, Oxford Series on Synchrotron Radiation (Oxford University Press, Oxford), ISBN 9780198501688.

Cormack, A. M. (1963). Representation of a function by its line integrals with some radiological applications, *Journal of Applied Physics* **34**, 2722–2727.

Cormack, A. M. (1964). Representation of a function by its line integrals, with some radiological applications. ii, *Journal of Applied Physics* **35**(10), 2908–2913.

Cormack, A. M. (1981). The Radon transform on a family of curves in the plane, *Proceedings of the American Mathematical Society* **83**(2), 325–330.

Cormack, A. M. (1982). The Radon transform on a family of curves in the plane. ii, *Proceedings of the American Mathematical Society* **86**(2), 293–298.

Cormack, A. M. (1984). Radon's problem – old and new, *SIAM-AMS Proceedings* **4**, 33–39.

Cormack, A. M. and Quinto, E. T. (1980). A Radon transform on spheres through the origin in $\mathbb{R}^n$ and applications to the Darboux equation, *Transactions of the American Mathematical Society* **260**(2), 575–581.

Cree, M. J. and Bones, P. J. (1994). Towards direct reconstruction from a gamma camera based on Compton scattering, *IEEE Transactions on Medical Imaging* **13**(2), 398–407.

Cruvinel, P. E. and Balogun, F. A. (2006). Compton scattering tomography for agricultural measurements, *Engenharia Agricola* **26**, 151–160.

Dahlbom, M. (ed.) (2017). *Physics of PET and SPECT Imaging* (CRC Press, Boca Raton, FL).

de Hoop, M., Saksala, T., Zhai, J., *et al.* (2019). Mixed ray transform on simple 2-dimensional Riemannian manifolds, *Proceedings of The American Mathematical Society* **147**(11), 4901–4913.

Denisjuk, A. (2006). Inversion of the x-ray transform for 3D symmetric tensor fields with sources on a curve, *Inverse Problems* **22**(2), 399.

Derevtsov, E. Y. and Pickalov, V. V. (2011). Reconstruction of vector fields and their singularities from ray transforms, *Numerical Analysis and Applications* **4**(1), 21–35.

Derevtsov, E. Y. and Svetov, I. (2015). Tomography of tensor fields in the plain, *Eurasian Journal of Mathematical and Computer Applications* **3**(2), 24–68.

Desai, N. M. and Lionheart, W. R. (2016). An explicit reconstruction algorithm for the transverse ray transform of a second rank tensor field from three axis data, *Inverse Problems* **32**(11), 115009.

Dogan, N., Wehe, D., and Akcasu, A. (1992). A source reconstruction method for multiple scatter Compton cameras, *IEEE Transactions on Nuclear Science* **39**(5), 1427–1430.

Dogan, N., Wehe, D. K., and Knoll, G. F. (1990). Multiple Compton scattering gamma ray imaging camera, *Nuclear Instruments and Methods in Physics Research Section A: Accelerators, Spectrometers, Detectors and Associated Equipment* **299**(1–3), 501–506.

Dohring, W., Reiss, K., and Fabel, H. (1974). Compton scatter for local in-vivo assessment of density in the lung, *Pneumonologie* **150**, 2-4, 345–359.

Drake, G. W. F. (ed.) (2006). *Springer Handbook of Atomic, Molecular, and Optical Physics* (Springer, New York).

Du, Y., He, Z., Knoll, G., Wehe, D., and Li, W. (2001). Evaluation of a Compton scattering camera using 3-D position sensitive CdZnTe detectors, *Nuclear Instruments and Methods in Physics Research Section A: Accelerators, Spectrometers, Detectors and Associated Equipment* **457**(1), 203–211.

Duderstadt, J. and Martin, W. (1979). *Transport Theory* (Wiley-Interscience Publications, New York).

Earnhart, J., Prettyman, T., Lestone, J., and Gardner, R. (2000). Simulation of Compton camera imaging with a specific purpose Monte Carlo code, *Applied Radiation and Isotopes* **53**(4-5), 673–680.

Earnhart, J. R. D. (1999). *A Compton Camera for Spectroscopic Imaging from 100keV to 1MeV*, Ph.D. Thesis, North Carolina State University, Raleigh, NC, USA, PhD Advisors: R. Gardner and T. Prettyman.

Ehrenpreis, L. (2003). *The Universality of the Radon Transform* (Oxford University Press).

Eskin, G. (2004). Inverse boundary value problems in domains with several obstacles, *Inverse Problems* **20**(5), 1497.

Evans, B. L., Martin, J. B., and Roggemann, M. C. (1999). Deconvolution of shift-variant broadening for Compton scatter imaging, *Nuclear Instruments and Methods in Physics Research Section A: Accelerators, Spectrometers, Detectors and Associated Equipment* **422**(1–3), 661–666.

Evans, B. L., Martin, J., Burggraf, L., and Roggemann, M. (1998). Nondestructive inspection using Compton scatter tomography, *IEEE Transactions on Nuclear Science* **45**(3), 950–956.

Evans, B. L., Martin, J. B., Burggraf, L. W., Roggemann, M., and Hangartner, T. (2002). Demonstration of energy-coded Compton scatter tomography with fan beams for one-sided inspection, *Nuclear Instruments and Methods in Physics Research Section A: Accelerators, Spectrometers, Detectors and Associated Equipment* **480**(2–3), pp. 797–806.

Everett, D. B., Fleming, J. S., Todd, R. W., and Nightingale, J. M. (1977). Gamma-radiation imaging system based on the Compton effect, *Proceedings of the Institution of Electrical Engineers* **124**, 11, 995–1000.

Farmer, F. and Collins, M. P. (1971). A new approach to the determination of anatomical cross-sections of the body by Compton scattering of gamma-rays, *Physics in Medicine & Biology* **16**(4), p. 577.

Farmer, F. and Collins, M. P. (1974). A further appraisal of the Compton scattering method for determining anatomical cross-sections of the body, *Physics in Medicine & Biology* **19**(6), 808.

Feeman, T. G. (2015). *The Mathematics of Medical Imaging: A Beginner's Guide* (Springer, Cham, Switzerland).

Felea, R. and Quinto, E. T. (2011). The microlocal properties of the local 3-D SPECT operator, *SIAM Journal on Mathematical Analysis* **43**(3), 1145–1157.

Fernández, J. (1999). Polarisation effects in multiple scattering photon calculations using the Boltzmann vector equation, *Radiation Physics and Chemistry* **56**(1–2), 27–59.

Fernández, J. E., Hubbell, J., Hanson, A., and Spencer, L. (1993). Polarization effects on multiple scattering gamma transport, *Radiation Physics and Chemistry* **41**(4–5), 579–630.

Finch, D. and Rakesh (2006). The range of the spherical mean value operator for functions supported in a ball, *Inverse Problems* **22**(3), 923.

Fischer, G. (2001). *Plane Algebraic Curves* (American Mathematical Society, Providence, RI).

Florescu, L., Markel, V. A., and Schotland, J. C. (2010). Single-scattering optical tomography: Simultaneous reconstruction of scattering and absorption, *Physical Review E* **81**, p. 016602.

Florescu, L., Markel, V. A., and Schotland, J. C. (2011). Inversion formulas for the broken-ray Radon transform, *Inverse Problems* **27**(2), 025002.

Florescu, L., Markel, V. A., and Schotland, J. C. (2018). Nonreciprocal broken ray transforms with applications to fluorescence imaging, *Inverse Problems* **34**(9), 094002.

Florescu, L., Schotland, J. C., and Markel, V. A. (2009). Single scattering optical tomography, *Physical Review E* **79**, p. 036607.

Folsom, M. (2020). *A compact neutron scatter camera using optical coded-aperture imaging*, Ph.D. thesis, University of Tennessee, Knoxville, PhD Advisor: J. Hayward.

Fontana, M., Dauvergne, D., Létang, J. M., Ley, J.-L., and Testa, É. (2017). Compton camera study for high efficiency SPECT and benchmark with Anger system, *Physics in Medicine & Biology* **62**(23), 8794.

Fouque, J.-P., Garnier, J., Papanicolaou, G., and Solna, K. (2007). *Wave Propagation and Time Reversal in Randomly Layered Media*, Vol. 56 (Springer New York, NY).

Frandes, M., Timar, B., and Lungeanu, D. (2016). Image reconstruction techniques for Compton scattering based imaging: an overview, *Current Medical Imaging Reviews* **12**, 1–1.

Frandes, M., Zoglauer, A., Maxim, V., and Prost, R. (2010). A tracking Compton-scattering imaging system for hadron therapy monitoring, *IEEE Transactions on Nuclear Science* **57**(1), 144–150.

Fremlin, D. H. (1974). *Topological Riesz Spaces and Measure Theory* (Cambridge University Press, Cambridge, UK).

Frikel, J. and Quinto, E. T. (2013). Characterization and reduction of artifacts in limited angle tomography, *Inverse Problems* **29**(12), p. 125007.

Gardner, R. J. (2006). *Geometric Tomography*, 2nd edn., Encyclopedia of Mathematics and its Applications (Cambridge University Press, Cambridge, UK).

Garnett, E., Kennett, T., Kenyon, D., and Webber, C. (1973). A photon scattering technique for the measurement of absolute bone density in man, *Radiology* **106**(1), 209–212.

Gautam, S., Hopkins, F., Klinksiek, R., and Morgan, I. (1983). Compton interaction tomography I. Feasibility studies for applications in earthquake engineering, *IEEE Transactions on Nuclear Science* **30**(2), 1680–1684.

Gelfand, I. M. and Shilov, G. E. (1964). *Generalized Functions*, Vol. 1: Properties and Operations (Academic Press, New York).

Gelfand, I. M., Gindikin, S. G., and Graev, M. I. (2003). *Selected Topics in Integral Geometry*, Translations of Mathematical Monographs (American Mathemaical Society, Providence, RI).

Gelfand, I. M., Graev, M. I., and Vilenkin, N. Y. (1966). *Generalized Functions*, Vol. 5: Integral Geometry and Representation Theory (Academic Press, New York, USA).

Goldsmith, J. E., Gerling, M. D., and Brennan, J. S. (2016). A compact neutron scatter camera for field deployment, *Review of Scientific Instruments* **87**(8), 083307.

Gormley, J., Rogers, W., Clinthorne, N., Wehe, D., Knell, G., and Wilderman, S. (1996). Experimental comparison of mechanical and electronic collimation, in *1996 IEEE Nuclear Science Symposium. Conference Record*, Vol. 2 (IEEE), 798–802.

Gorshkov, V., Kroening, M., Anosov, Y., and Dorjgochoo, O. (2005). X-ray scattering tomography, *Nondestructive Testing and Evaluation* **20**(3), 147–157.

Gouia-Zarrad, R. (2014). Analytical reconstruction formula for n-dimensional conical Radon transform, *Computers & Mathematics with Applications* **68**(9), 1016–1023.

Gouia-Zarrad, R. and Ambartsoumian, G. (2014). Exact inversion of the conical Radon transform with a fixed opening angle, *Inverse Problems* **30**(4), 045007.

Gouia-Zarrad, R. and Moon, S. (2018). Inversion of the attenuated conical Radon transform with a fixed opening angle, *Mathematical Methods in the Applied Sciences* **41**(18), 8423–8431.

Gradshteyn, I. S. and Ryzhik, I. M. (2007). *Table of Integrals, Series, and Products* (Academic Press, Amsterdam).

Griesmaier, R., Mishra, R. K., and Schmiedecke, C. (2018). Inverse source problems for Maxwell's equations and the windowed Fourier transform, *SIAM Journal on Scientific Computing* **40**(2), A1204–A1223.

Grubsky, V., Romanov, V., Patton, N., and Jannson, T. (2011). Compton imaging tomography technique for NDE of large nonuniform structures, in *Penetrating Radiation Systems and Applications XII*, Vol. 8144 (International Society for Optics and Photonics), p. 81440G.

Guzzardi, R. and Licitra, G. (1987). A critical review of Compton imaging. *Critical Reviews in Biomedical Engineering* **15**, 3, 237–268.

Haltmeier, M. (2014). Exact reconstruction formulas for a Radon transform over cones, *Inverse Problems* **30**(3), 035001.

Haltmeier, M., Moon, S., and Schiefeneder, D. (2017). Inversion of the attenuated V-line transform with vertices on the circle, *IEEE Transactions on Computational Imaging* **3**(4), 853–863.

Haltmeier, M. and Schiefeneder, D. (2018). Variational regularization of the weighted conical Radon transform, *Inverse Problems* **34**(12), 124009.

Han, L., Rogers, W. L., Huh, S. S., and Clinthorne, N. (2008). Statistical performance evaluation and comparison of a Compton medical imaging system and a collimated Anger camera for higher energy photon imaging, *Physics in Medicine & Biology* **53**(24), 7029.

Harding, G. (1982). On the sensitivity and application possibilities of a novel Compton scatter imaging system, *IEEE Transactions on Nuclear Science* **29**(3), 1259–1265.

Harding, G. (1997). Inelastic photon scattering: effects and applications in biomedical science and industry, *Radiation Physics and Chemistry* **50**(1), 91–111.

Harding, G. and Harding, E. (2010). Compton scatter imaging: A tool for historical exploration, *Applied Radiation and Isotopes* **68**(6), 993–1005.

Harding, G. and Kosanetzky, J. (1989). Scattered X-ray beam nondestructive testing, *Nuclear Instruments and Methods in Physics Research Section A: Accelerators, Spectrometers, Detectors and Associated Equipment* **280**(2–3), 517–528.

Harding, G. and Tischler, R. (1986). Dual-energy Compton scatter tomography, *Physics in Medicine & Biology* **31**(5), 477.

Harding, G., Strecker, H., and Tischler, R. (1983). X-ray imaging with Compton-scatter radiation, *Philips Technical Review* **41**(2), 46–59.

Hebert, T., Leahy, R., and Singh, M. (1990). Three-dimensional maximum-likelihood reconstruction for an electronically collimated single-photon-emission imaging system, *Journal of the Optical Society of America A* **7**(7), 1305–1313.

Helgason, S. (1999). *The Radon Transforms* (Birkhauser, Boston).

Helgason, S. (2011). *Integral Geometry and Radon Transforms* (Springer, New York).

Henyey, L. C. and Greenstein, J. L. (1940). Diffuse radiation in the galaxy, *The Astrophysical Journal* **93**, 70–83.

Hertle, A. (1988). The identification problem for the constantly attenuated Radon transform, *Mathematische Zeitschrift* **197**(1), 13–19.

Herzo, D., Koga, R., Millard, W., Moon, S., Ryan, J., Wilson, R., Zych, A., and White, R. (1975). A large double scatter telescope for gamma rays and neutrons, *Nuclear Instruments and Methods* **123**(3), 583–597.

Hilaire, E., Sarrut, D., Peyrin, F., and Maxim, V. (2016). Proton therapy monitoring by Compton imaging: influence of the large energy spectrum of the prompt-γ radiation, *Physics in Medicine & Biology* **61**(8), 3127.

Hill, W. H. and Matthews, K. L. (2007). Experimental verification of a hand held electronically-collimated radiation detector, in *2007 IEEE Nuclear Science Symposium Conference Record*, Vol. 5 (IEEE), 3792–3797.

Hirasawa, M. and Tomitani, T. (2003). An analytical image reconstruction algorithm to compensate for scattering angle broadening in Compton cameras, *Physics in Medicine & Biology* **48**(8), 1009–1026.

Hirasawa, M., Tomitani, T., and Shibata, S. (2001). New analytical method for three dimensional image reconstruction in multitracer gamma-ray emission imaging: Compton camera for multitracer, *RIKEN Review*, 118–119.

Holman, S. (2013). Generic local uniqueness and stability in polarization tomography, *Journal of Geometric Analysis* **23**(1), 229–269.

Holmes, R. B. (1975). *Geometric Functional Analysis and Its Applications* (Springer-Verlag, New York).

Holt, R. (1985). Compton imaging, *Endeavour* **9**(2), 97–105.

Holt, R. and Cooper, M. (1987). Gamma-ray scattering NDE, *NDT International* **20**(3), 161–165.

Holt, R. and Cooper, M. (1988). Non-destructive examination with a Compton scanner, *British Journal of Non-Destructive Testing* **30**(2), 75–80.

Holt, R., Cooper, M., and Jackson, D. (1984). Gamma-ray scattering techniques for non-destructive testing and imaging, *Nuclear Instruments and Methods in Physics Research* **221**(1), 98–104.

Holt, R., Kouris, K., Cooper, M., and Jackson, D. (1983). Assessment of gamma ray scattering for the characterisation of biological material, *Physics in Medicine & Biology* **28**(12), 1435.

Hörmander, L. (2015). *The Analysis of Linear Partial Differential Operators I: Distribution Theory and Fourier Analysis* (Springer, Berlin, Heidelberg).

Hristova, Y. (2015). Inversion of a V-line transform arising in emission tomography, *Journal of Coupled Systems and Multiscale Dynamics* **3**(3), 272–277.

Hruska, C. B., Phillips, S. W., Whaley, D. H., Rhodes, D. J., and O'connor, M. K. (2008). Molecular breast imaging: use of a dual-head dedicated gamma camera to detect small breast tumors, *American Journal of Roentgenology* **191**(6), 1805–1815.

Hua, C. (2000). *Compton Imaging System Development and Performance Assessment*, Ph.D. thesis, University of Michigan, PhD Advisor: W. L. Rogers.

Hua, C., Clinthorne, N., Wilderman, S., LeBlanc, J., and Rogers, W. (1999). Quantitative evaluation of information loss for Compton cameras, *IEEE Transactions on Nuclear Science* **46**(3), 587–593.

Hubenthal, M. (2014). The broken ray transform on the square, *Journal of Fourier Analysis and Applications* **20**(5), 1050–1082.

Hubenthal, M. (2015). The broken ray transform in n dimensions with flat reflecting boundary, *Inverse Problems & Imaging* **9**, 1, 143–161.

Huddleston, A. and Bhaduri, D. (1979). Compton scatter densitometry in cancellous bone, *Physics in Medicine & Biology* **24**(2), 310.

Hussein, E., Meneley, D., and Banerjee, S. (1986). On the solution of the inverse problem of radiation scattering imaging, *Nuclear Science and Engineering* **92**(3), 341–349.

Hussein, E. M. (2007). On the intricacy of imaging with incoherently-scattered radiation, *Nuclear Instruments and Methods in Physics Research Section B: Beam Interactions with Materials and Atoms* **263**(1), 27–31.

Hussein, E. M. and Waller, E. J. (1998). Review of one-side approaches to radiographic imaging for detection of explosives and narcotics, *Radiation Measurements* **29**(6), 581–591.

Hussein, E. M., Desrosiers, M., and Waller, E. J. (2005). On the use of radiation scattering for the detection of landmines, *Radiation Physics and Chemistry* **73**(1), 7–19.

Hutcheson, A. L. and Phlips, B. F. (2009). A liquid scintillator fast neutron double-scatter imager, in *2009 IEEE Nuclear Science Symposium Conference Record (NSS/MIC)* (IEEE), 1126–1128.

Hutton, B. F., Buvat, I., and Beekman, F. J. (2011). Review and current status of SPECT scatter correction, *Physics in Medicine & Biology* **56**(14), R85.

Ichihara, T. (1987). Ring type single-photon emission CT imaging apparatus, US Patent 4,639,599, January 27.

Ilmavirta, J. (2013). Broken ray tomography in the disc, *Inverse Problems* **29**(3), 035008.

Ilmavirta, J. and Paternain, G. P. (2018). Broken ray tensor tomography with one reflecting obstacle, *arXiv preprint arXiv:1805.04947* .

Ilmavirta, J. and Salo, M. (2016). Broken ray transform on a Riemann surface with a convex obstacle, *Communications in Analysis and Geometry* **24**(2), 379–408.

Ishimaru, A. (2017). *Electromagnetic Wave Propagation, Radiation, and Scattering: From Fundamentals to Applications*, IEEE Press Series on Electromagnetic Wave Theory (Wiley, Piscataway, NJ).

Jacobs, A. M. (1986). Compton profile radiography, *International Advances in Nondestructive Testing* **12**, pp. 17–52.

Jeon, G. and Moon, S. (2021). Singular value decomposition of the attenuated conical Radon transform with a fixed central axis and opening angle, *Integral Transforms and Special Functions* **32**(10), 812–822.

Jha, A. K., Kupinski, M. A., Masumura, T., Clarkson, E., Maslov, A. V., and Barrett, H. H. (2012). Simulating photon-transport in uniform media using the radiative transport equation: a study using the Neumann-series approach, *Journal of the Optical Society of America A* **29**(8), 1741–1757.

Jones, K. C., Redler, G., Templeton, A., Bernard, D., Turian, J. V., and Chu, J. C. H. (2018). Characterization of Compton-scatter imaging with an analytical simulation method, *Physics in Medicine & Biology* **63**(2), 025016.

Jung, C.-Y. and Moon, S. (2015). Inversion formulas for cone transforms arising in application of Compton cameras, *Inverse Problems* **31**(1), 015006.

Jung, C.-Y. and Moon, S. (2016). Exact inversion of the cone transform arising in an application of a Compton camera consisting of line detectors, *SIAM Journal on Imaging Sciences* **9**(2), 520–536.

Kamae, T., Enomoto, R., and Hanada, N. (1987). A new method to measure energy, direction, and polarization of gamma rays, *Nuclear Instruments and Methods in Physics Research Section A: Accelerators, Spectrometers, Detectors and Associated Equipment* **260**(1), 254–257.

Kamae, T., Hanada, N., and Enomoto, R. (1988). Prototype design of multiple Compton gamma-ray camera, *IEEE Transactions on Nuclear Science* **35**(1), 352–355.

Katsevich, A. and Krylov, R. (2013). Broken ray transform: inversion and a range condition, *Inverse Problems* **29**(7), 075008.

Katsevich, A. and Schuster, T. (2013). An exact inversion formula for cone beam vector tomography, *Inverse Problems* **29**(6), 065013.

Kazantsev, S. and Bukhgeim, A. (2004). Singular value decomposition for the 2D fan-beam Radon transform of tensor fields, *Journal of Inverse and Ill-Posed Problems* **12**(3), 245–278.

Kim, D. and Wongsason, P. (2020). Three-dimensional vector field inversion formula using first moment transverse transform in quaternionic approaches, *Mathematical Methods in the Applied Sciences* **43**(12), 7070–7086.

Kim, N. and Lee, C. (2008). Monte Carlo simulation study on medical imaging of a multiply stacked Compton camera, *Journal of the Korean Physical Society* **53**, 1201–1204.

Kim, S. M. (2018). Analytic simulator and image generator of multiple-scattering Compton camera for prompt gamma ray imaging, *Biomedical Engineering Letters* **8**(4), 383–392.

Kim, S. M., Lee, J. S., Lee, M. N., Lee, J. H., Lee, C. S., Kim, C.-H., Lee, D. S., and Lee, S.-J. (2007). Two approaches to implementing projector-backprojector pairs for 3D reconstruction from Compton scattered data, *Nuclear Instruments and Methods in Physics Research Section A: Accelerators, Spectrometers, Detectors and Associated Equipment* **571**(1–2), 255–258.

King, S., Phillips, G., Haskins, P., McKisson, J., Piercey, R., and Mania, R. (1994). A solid-state Compton camera for three-dimensional imaging, *Nuclear Instruments and Methods in Physics Research Section A: Accelerators, Spectrometers, Detectors and Associated Equipment* **353**(1–3), 320–323.

Kishimoto, A., Kataoka, J., Taya, T., Tagawa, L., Mochizuki, S., Ohsuka, S., Nagao, Y., Kurita, K., Yamaguchi, M., Kawachi, N., *et al.* (2017). First demonstration of multi-color 3-D in vivo imaging using ultra-compact Compton camera, *Scientific Reports* **7**(1), 1–7.

Kolkoori, S., Wrobel, N., Zscherpel, U., and Ewert, U. (2015). A new X-ray backscatter imaging technique for non-destructive testing of aerospace materials, *NDT&E International* **70**, 41–52.

Kondic, N. (1978). Density field determination by an external stationary radiation source using a kernel technique, in *Measuremnts in Polyphase Flows Symposium* (American Society of Mechanical Engineers, San Francisco California, USA), 37–51.

Kondic, N. and Hahn, O. J. (1970). Theory and application of the parallel and diverging radiation beam method in two-phase systems, in *International Heat Transfer Conference 4* (Elsevier, Paris-Versailles, France), 1–10.

Kondic, N., Jacobs, A., and Ebert, D. (1983). Three-dimensional density field determination by external stationary detectors and gamma sources using selective scattering, in *Proceedings of 2-nd International Topical Meeting on Nuclear Reactor Thermal Hydraulics*, Vol. II (American Nuclear Society, La Grange Park, IL, USA), 1443–1455.

Krane, K. S. (2019). *Modern Physics*. 4th edn. (Wiley, New York).

Krishnan, V., Manna, R., Sahoo, S., and Sharafutdinov, V. (2019a). Momentum ray transforms, *Inverse Problems & Imaging* **13**(3), 679–701.

Krishnan, V. P. and Mishra, R. K. (2018). Microlocal analysis of a restricted ray transform on symmetric m-tensor fields in $\mathbb{R}^n$, *SIAM Journal on Mathematical Analysis* **50**(6), 6230–6254.

Krishnan, V. P., Mishra, R. K., and Monard, F. (2019b). On solenoidal-injective and injective ray transforms of tensor fields on surfaces, *Journal of Inverse and Ill-posed Problems* **27**(4), 527–538.

Krishnan, V. P., Mishra, R. K., and Sahoo, S. K. (2021). Microlocal inversion of a 3-dimensional restricted transverse ray transform on symmetric tensor fields, *Journal of Mathematical Analysis and Applications* **495**(1), 124700.

Krishnan, V. P. and Quinto, E. T. (2015). Microlocal analysis in tomography, in O. Scherzer (ed.), *Handbook of Mathematical Methods in Imaging* (Springer New York, New York, NY), 847–902.

Krishnan, V. P. and Sharafutdinov, V. A. (2022). Ray transform on Sobolev spaces of symmetric tensor fields, I: Higher order Reshetnyak formulas, *Inverse Problems & Imaging* **16**(4), 787–826.

Krishnan, V. P. and Stefanov, P. (2009). A support theorem for the geodesic ray transform of symmetric tensor fields, *Inverse Problems & Imaging* **3**(3), 453.

Krishnan, V. P., Manna, R., Sahoo, S. K., and Sharafutdinov, V. A. (2020). Momentum ray transforms, ii: range characterization in the Schwartz space, *Inverse Problems* **36**(4), 045009.

Kroeger, R., Johnson, W., Kurfess, J., Phlips, B., and Wulf, E. (2002). Three-Compton telescope: theory, simulations, and performance, *IEEE Transactions on Nuclear Science* **49**(4), 1887–1892.

Krylov, R. and Katsevich, A. (2015). Inversion of the broken ray transform in the case of energy dependent attenuation, *Physics in Medicine & Biology* **60**(11), 4313–4334.

Kuchment, P. (2013). *The Radon Transform and Medical Imaging* (SIAM, Philadelphia, PA).

Kuchment, P. and Terzioglu, F. (2016). Three-dimensional image reconstruction from Compton camera data, *SIAM Journal on Imaging Sciences* **9**(4), 1708–1725.

Kuchment, P. and Terzioglu, F. (2017). Inversion of weighted divergent beam and cone transforms, *Inverse Problems & Imaging* **11**(6), 1071–1090.

Kuger, L. and Rigaud, G. (2020). Joint fan-beam CT and Compton scattering tomography: analysis and image reconstruction, *arXiv preprint arXiv:2008.06699*.

Kunyansky, L. A. (2001). A new SPECT reconstruction algorithm based on the novikov explicit inversion formula, *Inverse Problems* **17**(2), 293.

Kurfess, J. D., Johnson, W., Kroeger, R., Novikova, E., Phlips, B., Strickman, M., and Wulf, E. (2004). An advanced Compton telescope based on thick, position-sensitive solid-state detectors, *New Astronomy Reviews* **48**(1–4), 293–298.

Kurosawa, S., Kubo, H., Ueno, K., Kabuki, S., Iwaki, S., Takahashi, M., Taniue, K., Higashi, N., Miuchi, K., Tanimori, T., *et al.* (2012). Prompt gamma detection for range verification in proton therapy, *Current Applied Physics* **12**(2), 364–368.

Kurylev, Y., Lassas, M., and Uhlmann, G. (2010). Rigidity of broken geodesic flow and inverse problems, *American Journal of Mathematics* **132**(2), 529–562.

Kwon, K. (2019). An inversion of the conical Radon transform arising in the Compton camera with helical movement, *Biomedical Engineering Letters* **9**(2), 233–243.

Lackie, A. W., Matthews, K. L., Smith, B. M., Hill, W., Wang, W.-H., and Cherry, M. L. (2006). A directional algorithm for an electronically-collimated gamma-ray detector, in *2006 IEEE Nuclear Science Symposium Conference Record*, Vol. 1 (IEEE), 264–269.

Lale, P. G. (1959). The examination of internal tissues, using gamma-ray scatter with a possible extension to megavoltage radiography, *Physics in Medicine & Biology* **4**(2), 159–167.

Lale, P. G. (1968). The examination of internal tissues by high-energy scattered X radiation, *Radiology* **90**(3), 510–517.

Lebesgue, H. (1910). Sur l'intégration des fonctions discontinues, *Annales scientifiques de l'École Normale Supérieure* **27**, p. 361–450.

Leblanc, J. W. (1999). *A Compton Camera for Low Energy Gamma Ray Imaging in Nuclear Medicine Applications*, Ph.D. thesis, University of Michigan, PhD Advisor: D. K. Wehe.

LeBlanc, J., Clinthorne, N., Hua, C.-H., Nygard, E., Rogers, W., Wehe, D., Weilhammer, P., and Wilderman, S. (1998). C-SPRINT: a prototype Compton camera system for low energy gamma ray imaging, *IEEE Transactions on Nuclear Science* **45**(3), 943–949.

LeBlanc, J., Clinthorne, N., Hua, C., Rogers, W., Wehe, D., and Wilderman, S. (1999a). A Compton camera for nuclear medicine applications using ^{113m}In, *Nuclear Instruments and Methods in Physics Research Section A: Accelerators, Spectrometers, Detectors and Associated Equipment* **422**(1–3), 735–739.

LeBlanc, J., Clinthorne, N., Hua, C.-H., Nygard, E., Rogers, W., Wehe, D., Weilhammer, P., and Wilderman, S. (1999b). Experimental results from the C-SPRINT prototype Compton camera, *IEEE Transactions on Nuclear Science* **46**(3), 201–204.

Lee, J. (2012). *Introduction to Smooth Manifolds, Graduate Texts in Mathematics*, Vol. 218, 2nd edn. (Springer).

Lee, J.-S. and Chen, J.-C. (2015). A single scatter model for X-ray CT energy spectrum estimation and polychromatic reconstruction, *IEEE Transactions on Medical Imaging* **34**(6), 1403–1413.

Lee, W. and Lee, T. (2010). A compact Compton camera using scintillators for the investigation of nuclear materials, *Nuclear Instruments Methods in Physics Research Section A, Accelerators, Spectrometers, Detectors and Associated Equipment* **624**(1), 118–124.

Lewis, R. A., Rogers, K., Hall, C. J., Towns-Andrews, E., Slawson, S., Evans, A., Pinder, S., Ellis, I., Boggis, C., Hufton, A. P., *et al.* (2000). Breast cancer diagnosis using scattered X-rays, *Journal of Synchrotron Radiation* **7**(5), 348–352.

Li, J., Valentine, J. D., Aarsvold, J. N., and Khamzin, M. (2001). A rebinning technique for 3D reconstruction of Compton camera data, in *2001 IEEE Nuclear Science Symposium Conference Record (Cat. No. 01CH37310)*, Vol. 4 (IEEE), 1877–1881.

Lingenfelter, D. J., Fessler, J. A., Scott, C. D., and He, Z. (2010). Benefits of position-sensitive detectors for radioactive source detection, *IEEE Transactions on Signal Processing* **58**(9), 4473–4483.

Lvin, S. (1994). Data correction and restoration in emission tomography, *Tomography, Impedance Imaging, and Integral Geometry, Lectures in Application Mathematics* **30**, 149–155.

Macdonald, I. G. (2015). *Symmetric Functions and Hall Polynomials* (Oxford University Press, Oxford, UK).

Martin, J., Dogan, N., Gormley, J., Knoll, G., O'Donnell, M., and Wehe, D. (1994). Imaging multi-energy gamma-ray fields with a Compton scatter camera, *IEEE Transactions on Nuclear Science* **41**(4), 1019–1025.

Martin, J., Knoll, G., Wehe, D., Dogan, N., Jordanov, V., Petrick, N., and Singh, M. (1993). A ring Compton scatter camera for imaging medium energy gamma rays, *IEEE Transactions on Nuclear Science* **40**(4), 972–978.

Masahiro, E., Takanori, T., Nobuyuki, N., and Katsuya, Y. (2001). Effect of scattered radiation on image noise in cone beam CT, *Medical Physics* **28**(4), 469–474.

MATLAB (2020). *Head Phantom Image* (The MathWorks, Inc., Natick, MA), https://www.mathworks.com/help/images/ref/phantom.html.

Maxim, V. (2014). Filtered backprojection reconstruction and redundancy in Compton camera imaging, *IEEE Transactions on Image Processing* **23**(1), 332–341.

Maxim, V. (2019). Enhancement of Compton camera images reconstructed by inversion of a conical Radon transform, *Inverse Problems* **35**(1), 014001.

Maxim, V., Frandeş, M., and Prost, R. (2009). Analytical inversion of the Compton transform using the full set of available projections, *Inverse Problems* **25**(9), 095001.

Maxim, V., Lojacono, X., Hilaire, E., Krimmer, J., Testa, E., Dauvergne, D., Magnin, I., and Prost, R. (2015). Probabilistic models and numerical calculation of system matrix and sensitivity in list-mode MLEM 3D reconstruction of Compton camera images, *Physics in Medicine & Biology* **61**(1), 243.

McKisson, J., Haskins, P., Phillips, G., King, S., August, R., Piercey, R., and Mania, R. (1994). Demonstration of three-dimensional imaging with a germanium Compton camera, *IEEE Transactions on Nuclear Science* **41**(4), 1221–1224.

Mennessier, C., Noo, F., Clackdoyle, R., Bal, G., and Desbat, L. (1999). Attenuation correction in SPECT using consistency conditions for the exponential ray transform, *Physics in Medicine & Biology* **44**(10), 2483.

Mishra, R. K. (2020). Full reconstruction of a vector field from restricted Doppler and first integral moment transforms in $\mathbb{R}^n$, *Journal of Inverse and Ill-posed Problems* **28**(2), 173–184.

Mishra, R. K. and Sahoo, S. K. (2021). Injectivity and range description of integral moment transforms over m-tensor fields in $\mathbb{R}^n$, *SIAM Journal on Mathematical Analysis* **53**(1), 253–278.

Monard, F. (2016). Efficient tensor tomography in fan-beam coordinates, *Inverse Problems & Imaging* **10**(2), 433–459.

Moon, S. (2016). On the determination of a function from its conical Radon transform with fixed central axis, *SIAM Journal on Mathematical Analysis* **48**(3), 1833–1847.

Moon, S. (2017). Inversion of the conical Radon transform with vertices on a surface of revolution arising in an application of a Compton camera, *Inverse Problems* **33**(6), 065002.

Moon, S. (2019). Orthogonal function series formulae for inversion of the conical Radon transform with a fixed central axis, *Inverse Problems* **35**(12), 125007.

Moon, S. and Haltmeier, M. (2017). Analytic inversion of a conical Radon transform arising in application of Compton cameras on the cylinder, *SIAM Journal on Imaging Sciences* **10**(2), 535–557.

Moon, S. and Haltmeier, M. (2020). The conical Radon transform with vertices on triple line segments, *Inverse Problems* **36**, 11, p. 115005.

Morvidone, M., Nguyen, M. K., Truong, T. T., and Zaidi, H. (2010). On the V-line Radon transform and its imaging applications, *International Journal of Biomedical Imaging* **2010**, p. 208179.

Motomura, S., Kanayama, Y., Haba, H., Watanabe, Y., and Enomoto, S. (2008). Multiple molecular simultaneous imaging in a live mouse using semiconductor Compton camera, *Journal of Analytical Atomic Spectrometry* **23**(8), 1089–1092.

Muñoz, E., Barrio, J., Bemmerer, D., Etxebeste, A., Fiedler, F., Hueso-González, F., Lacasta, C., Oliver, J., Römer, K., Solaz, C., *et al.* (2018). Tests of MACACO Compton telescope with 4.44 MeV gamma rays, *Journal of Instrumentation* **13**(05), P05007.

Natterer, F. (1983). Exploiting the ranges of Radon transforms in tomography, in *Numerical Treatment of Inverse Problems in Differential and Integral Equations* (Springer), 290–303.

Natterer, F. (2001a). Inversion of the attenuated Radon transform, *Inverse Problems* **17**(1), 113.

Natterer, F. (2001b). *The Mathematics of Computerized Tomography*, Classics in Mathematics (Society for Industrial and Applied Mathematics, New York).

Natterer, F. and Wübbeling, F. (2001). *Mathematical Methods in Image Reconstruction*, Monographs on Mathematical Modeling and Computation (Society for Industrial and Applied Mathematics, New York).

Nguyen, L. V. (2015). How strong are streak artifacts in limited angle computed tomography? *Inverse Problems* **31**(5), 055003.

Nguyen, D. N. and Nguyen, L. V. (2021a). An inversion formula for the horizontal conical Radon transform, *Analysis and Mathematical Physics* **11**(1), 1–11.

Nguyen, D. N. and Nguyen, L. V. (2021b). Sampling for the V-line transform with vertex on a circle, *Inverse Problems* **37**(7), 075004.

Nguyen, M. K. and Truong, T. (2002). On an integral transform and its inverse in nuclear imaging, *Inverse Problems* **18**(1), 265.

Nguyen, M. K. and Truong, T. T. (2010). Inversion of a new circular-arc Radon transform for Compton scattering tomography, *Inverse Problems* **26**(6), 065005.

Nguyen, M. K., Truong, T. T., and Grangeat, P. (2005). Radon transforms on a class of cones with fixed axis direction, *Journal of Physics A: Mathematical and General* **38**(37), 8003.

Nguyen, M. K., Truong, T. T., Bui, H. D., and Delarbre, J. L. (2004). A novel inverse problem in γ-rays emission imaging, *Inverse Problems in Science and Engineering* **12**(2), 225–246.

Nguyen, M. K., Truong, T. T., Driol, C., and Zaidi, H. (2009). On a novel approach to Compton scattered emission imaging, *IEEE Transactions on Nuclear Science* **56**(3), 1430–1437.

Nguyen, M. K., Truong, T. T., Morvidone, M., and Zaidi, H. (2011). Scattered radiation emission imaging: principles and applications, *International Journal of Biomedical Imaging* **2011**, p. 913893.

Ning, R., Tang, X., and Conover, D. (2004). X-ray scatter correction algorithm for cone beam CT imaging, *Medical Physics* **31**(5), 1195–1202.

Norton, S. J. (1992). Unique tomographic reconstruction of vector fields using boundary data, *IEEE Transactions on Image Processing* **1**(3), 406–412.

Norton, S. J. (1994). Compton scattering tomography, *Journal of Applied Physics* **76**(4), 2007–2015.

Novikov, R. (2002a). An inversion formula for the attenuated X-ray transformation, *Arkiv för Matematik* **40**(1), 145–167.

Novikov, R. (2002b). On the range characterization for the two-dimensional attenuated x-ray transformation, *Inverse Problems* **18**(3), 677.

Novikov, R. and Sharafutdinov, V. (2007). On the problem of polarization tomography: I, *Inverse Problems* **23**(3), 1229.

Olson, A., Ciabatti, A., Hristova, Y., Kuchment, P., Ragusa, J., and Allmaras, M. (2016). Passive detection of small low-emission sources: two-dimensional numerical case studies, *Nuclear Science and Engineering* **184**(1), 125–150.

Ordonez, C. E., Bolozdynya, A., and Chang, W. (1997a). Dependence of angular uncertainties on the energy resolution of Compton cameras, in *1997 IEEE Nuclear Science Symposium Conference Record*, Vol. 2, 1122–1125.

Ordonez, C. E., Bolozdynya, A., and Chang, W. (1997b). Doppler broadening of energy spectra in Compton cameras, in *1997 IEEE Nuclear Science Symposium Conference Record*, Vol. 2, pp. 1361–1365.

Ordonez, C. E., Chang, W., and Bolozdynya, A. (1999). Angular uncertainties due to geometry and spatial resolution in Compton cameras, *IEEE Transactions on Nuclear Science* **46**(4), 1142–1147.

Palamodov, V. (2004). *Reconstructive Integral Geometry*, Monographs in Mathematics (Birkhaüser).

Palamodov, V. (2009). Reconstruction of a differential form from Doppler transform, *SIAM Journal on Mathematical Analysis* **41**(4), 1713–1720.

Palamodov, V. P. (2011). An analytic reconstruction for the Compton scattering tomography in a plane, *Inverse Problems* **27**, 12, p. 125004.

Palamodov, V. (2016). *Reconstruction from Integral Data*, Monographs and Research Notes in Mathematics (Chapman and Hall/CRC).

Palamodov, V. (2017). Reconstruction from cone integral transforms, *Inverse Problems* **33**(10), 104001.

Parra, L. C. (2000). Reconstruction of cone-beam projections from Compton scattered data, *IEEE Transactions on Nuclear Science* **47**(4), 1543–1550.

Patch, S. (2004). Thermoacoustic tomography — consistency conditions and the partial scan problem, *Physics in Medicine & Biology* **49**(11), 2305.

Paternain, G. P., Salo, M., and Uhlmann, G. (2013). Tensor tomography on surfaces, *Inventiones Mathematicae* **193**(1), 229–247.

Paternain, G. P., Salo, M., and Uhlmann, G. (2014). Tensor tomography: progress and challenges, *Chinese Annals of Mathematics (Series B)* **3**(35), 399—428.

Pauli, J., Pauli, E.-M., and Anton, G. (2002). ITEM—QM solutions for EM problems in image reconstruction exemplary for the Compton camera, *Nuclear Instruments and Methods in Physics Research Section A: Accelerators, Spectrometers, Detectors and Associated Equipment* **488**(1–2), 323–331.

Phillips, G. W. (1995). Gamma-ray imaging with Compton cameras, *Nuclear Instruments and Methods in Physics Research Section B: Beam Interactions with Materials and Atoms* **99**(1–4), 674–677.

Phillips, G. W. (1997). Applications of Compton imaging in nuclear waste characterization and treaty verification, in *1997 IEEE Nuclear Science Symposium Conference Record*, Vol. 1 (IEEE), 362–364.

Ponomaryov, I. (1995). Correction of emission tomography data: effects of detector displacement and non-constant sensitivity, *Inverse Problems* **11**(5), 1031.

Prado, P. G., Nguyen, M. K., Dumas, L., and Cohen, S. X. (2017). Three-dimensional imaging of flat natural and cultural heritage objects by a Compton scattering modality, *Journal of Electronic Imaging* **26**(1), 011026.

Prettyman, T., Gardner, R., Russ, J., and Verghese, K. (1993). A combined transmission and scattering tomographic approach to composition and density imaging, *Applied Radiation and Isotopes* **44**(10–11), 1327–1341.

Quinto, E. T. (1983). Singular value decompositions and inversion methods for the exterior Radon transform and a spherical transform, *Journal of mathematical Analysis and Applications* **95**(2), 437–448.

Ramaseshan, K. (2004). Microlocal analysis of the Doppler transform on $\mathbb{R}^3$, *Journal of Fourier Analysis and Applications* **10**(1), 73–82.

Ramlau, R. and Scherzer, O. (2019). *The Radon Transform: The First 100 Years and Beyond, Radon Series on Computational and Applied Mathematics*, Vol. 22 (Walter de Gruyter GmbH & Co KG, De Gruyter, Berlin).

Redler, G., Jones, K. C., Templeton, A., Bernard, D., Turian, J., and Chu, J. C. H. (2018). Compton scatter imaging: A promising modality for image guidance in lung stereotactic body radiation therapy, *Medical Physics* **45**(3), 1233–1240.

Reiss, K.-H. and Schuster, W. (1972). Quantitative measurements of lung function in children by means of Compton backscatter, *Radiology* **102**(3), 613–617.

Richard, M.-H., Chevallier, M., Dauvergne, D., Freud, N., Henriquet, P., Le Foulher, F., Letang, J., Montarou, G., Ray, C., Roellinghoff, F., *et al.* (2010). Design guidelines for a double scattering Compton camera for prompt-γ imaging during ion beam therapy: a Monte Carlo simulation study, *IEEE Transactions on Nuclear Science* **58**(1), 87–94.

Rigaud, G. (2013). On the inversion of the Radon transform on a generalized Cormack-type class of curves, *Inverse Problems* **29**(11), 115010.

Rigaud, G. (2017). Compton scattering tomography: feature reconstruction and rotation-free modality, *SIAM Journal on Imaging Sciences* **10**(4), 2217–2249.

Rigaud, G. (2021). 3D Compton scattering imaging with multiple scattering: Analysis by FIO and contour reconstruction, *Inverse Problems* **37**(6), 064001.

Rigaud, G. and Hahn, B. N. (2018). 3D Compton scattering imaging and contour reconstruction for a class of Radon transforms, *Inverse Problems* **34**(7), 075004.

Rigaud, G. and Hahn, B. N. (2021). Reconstruction algorithm for 3D Compton scattering imaging with incomplete data, *Inverse Problems in Science and Engineering* **29**(7), 967–989.

Rigaud, G., Nguyen, M. K., and Louis, A. K. (2012). Novel numerical inversions of two circular-arc Radon transforms in Compton scattering tomography, *Inverse Problems in Science and Engineering* **20**, 6, 809–839.

Rigaud, G., Régnier, R., Nguyen, M. K., and Zaidi, H. (2013). Combined modalities of Compton scattering tomography, *IEEE Transactions on Nuclear Science* **60**(3), 1570–1577.

Rogers, W., Clinthorne, N., and Bolozdynya, A. (2004). Compton cameras for nuclear medical imaging, in *Emission Tomography: The Fundamentals of PET and SPECT* (Academic Press), 383–419.

Rohe, R., Sharfi, M., Kecevar, K., Valentine, J., and Bonnerave, C. (1997). The spatially-variant backprojection point kernel function of an energy-subtraction Compton scatter camera for medical imaging, *IEEE Transactions on Nuclear Science* **44**(6), 2477–2482.

Rohe, R. C. and Valentine, J. D. (1996). An energy-subtraction Compton scatter camera design for in vivo medical imaging of radiopharmaceuticals, *IEEE Transactions on Nuclear Science* **43**(6), 3256–3263.

Royden, H. L. and Fitzpatrick, P. (2010). *Real Analysis* (Prentice Hall, Boston, USA).

Royle, G. and Speller, R. (1996). A flexible geometry Compton camera for industrial gamma ray imaging, in *1996 IEEE Nuclear Science Symposium. Conference Record*, Vol. 2 (IEEE), 821–824.

Royle, G. and Speller, R. (1997). Compton scatter imaging of a nuclear industry site, in *1997 IEEE Nuclear Science Symposium Conference Record*, Vol. 1 (IEEE), 365–368.

Rubin, B. (2015). *Introduction to Radon Transforms: With Elements of Fractional Calculus and Harmonic Analysis, Encyclopedia of Mathematics and its Applications*, Vol. 160 (Cambridge University Press, Cambridge, UK).

Rudin, W. (1973). *Functional Analysis* (McGraw-Hill, New York).

Rudin, W. (1987). *Real and Complex Analysis, 3rd Ed.* (McGraw-Hill, New York).

Sadiq, K. and Tamasan, A. (2015). On the range characterization of the two-dimensional attenuated Doppler transform, *SIAM Journal on Mathematical Analysis* **47**(3), 2001–2021.

Sadiq, K., Scherzer, O., and Tamasan, A. (2016). On the X-ray transform of planar symmetric 2-tensors, *Journal of Mathematical Analysis and Applications* **442**(1), 31–49.

Sauve, A. C., Hero, A. O., Rogers, W. L., Wilderman, S. J., and Clinthorne, N. H. (1999). 3D image reconstruction for a Compton SPECT camera model, *IEEE Transactions on Nuclear Science* **46**, 6, 2075–2084.

Scannavini, M., Speller, R., Royle, G., Cullum, I., Raymond, M., Hall, G., and Iles, G. (2000). Design of a small laboratory Compton camera for the imaging of positron emitters, *IEEE Transactions on Nuclear Science* **47**(3), 1155–1162.

Schiefeneder, D. and Haltmeier, M. (2017). The Radon transform over cones with vertices on the sphere and orthogonal axes, *SIAM Journal on Applied Mathematics* **77**(4), 1335–1351.

Schönfelder, V., Aarts, H., Bennett, K., Deboer, H., Clear, J., Collmar, W., Connors, A., Deerenberg, A., Diehl, R., Von Dordrecht, A., *et al.* (1993). Instrument description and performance of the imaging gamma-ray telescope COMPTEL aboard the Compton Gamma-Ray Observatory, *Astrophysical Journal Supplement Series* **86**, 657–692.

Schönfelder, V., Hirner, A., and Schneider, K. (1973). A telescope for soft gamma ray astronomy, *Nuclear Instruments and Methods* **107**(2), 385–394.

Schotland, J. C. (2012). Direct reconstruction methods in optical tomography, in H. Ammari (ed.), *Mathematical Modeling in Biomedical Imaging II: Optical, Ultrasound, and Opto-Acoustic Tomographies* (Springer, Berlin, Heidelberg), 1–29.

Schuster, T. (2000). The 3D Doppler transform: elementary properties and computation of reconstruction kernels, *Inverse Problems* **16**(3), 701.

Seo, H., Kim, C. H., Park, J. H., Kim, J. K., Lee, J. H., Lee, C. S., and Lee, J. S. (2010). Development of double-scattering-type Compton camera with double-sided silicon strip detectors and NaI (Tl) scintillation detector, *Nuclear Instruments and Methods in Physics Research Section A: Accelerators, Spectrometers, Detectors and Associated Equipment* **615**(3), 333–339.

Sethi, A. (2006). X-rays: Interaction with matter, in *Encyclopedia of Medical Devices and Instrumentation* (American Cancer Society), 590–599.

Sharaf, J. (2001). Practical aspects of Compton scatter densitometry, *Applied Radiation and Isotopes* **54**(5), 801–809.

Sharafutdinov, V. (2007). Slice-by-slice reconstruction algorithm for vector tomography with incomplete data, *Inverse Problems* **23**, 6, p. 2603.

Sharafutdinov, V. (2008). The problem of polarization tomography: II, *Inverse Problems* **24**(3), 035010.

Sharafutdinov, V. A. (1986). A problem of integral geometry for generalized tensor fields on $\mathbb{R}^n$, in *Doklady Akademii Nauk*, 2 (Russian Academy of Sciences), 305–307.

Sharafutdinov, V. A. (1994). *Integral Geometry of Tensor Fields* (Walter de Gruyter, De Gruyter, Berlin).

Sharafutdinov, V. A. (2016). The Reshetnyak formula and Natterer stability estimates in tensor tomography, *Inverse Problems* **33**(2), 025002.

Sherson, B. (2015). *Some Results in Single-Scattering Tomography*, Ph.D. thesis, Oregon State University, PhD Advisor: D. Finch.

Sinclair, L., Hanna, D., MacLeod, A., and Saull, P. (2009). Simulations of a scintillator Compton gamma imager for safety and security, *IEEE Transactions on Nuclear Science* **56**(3), 1262–1268.

Singh, M. (1983). An electronically collimated gamma camera for single photon emission computed tomography. Part I: Theoretical considerations and design criteria, *Medical Physics* **10**(4), 421–427.

Singh, M. and Brechner, R. R. (1990). Experimental test-object study of electronically collimated SPECT, *Journal of Nuclear Medicine* **31**(2), 178–186.

Singh, M. and Doria, D. (1983). An electronically collimated gamma camera for single photon emission computed tomography. Part II: Image reconstruction and preliminary experimental measurements, *Medical Physics* **10**(4), 428–435.

Singh, M., Doty, F., Friesenhahn, S., and Butler, J. (1995). Feasibility of using cadmium-zinc-telluride detectors in electronically collimated SPECT, *IEEE Transactions on Nuclear Science* **42**(4), 1139–1146.

Singh, M., Leahy, R., Brechner, R., and Hebert, T. (1988). Noise propagation in electronically collimated single photon imaging, *IEEE Transactions on Nuclear Science* **35**(1), 772–777.

Smith, B. D. (2005). Reconstruction methods and completeness conditions for two Compton data models, *Journal of the Optical Society of America A* **22**(3), 445–459.

Smith, B. D. (2011). Line-reconstruction from Compton cameras: data sets and a camera design, *Optical Engineering* **50**(5), 053204.

Smith, B. D. (2012). Computer simulations to demonstrate new inversion methods for Compton camera data, *Optical Engineering* **51**(5), p. 053203.

Smith, B. D. (2015). A new Compton camera imaging model to mitigate the finite spatial resolution of detectors and new camera designs for implementation, *Technologies* **3**(4), 219–237.

Solmon, D. C. (1995). The identification problem for the exponential Radon transform, *Mathematical Methods in the Applied Sciences* **18**(9), 687–695.

Solomon, C. J. and Ott, R. J. (1988). Gamma ray imaging with silicon detectors — a Compton camera for radionuclide imaging in medicine, *Nuclear Instruments and Methods in Physics Research Section A: Accelerators, Spectrometers, Detectors and Associated Equipment* **273**(2–3), 787–792.

Sparr, G., Strahlen, K., Lindstrom, K., and Persson, H. W. (1995). Doppler tomography for vector fields, *Inverse Problems* **11**(5), 1051.

Speller, R. and Horrocks, J. (1991). Photon scattering — a 'new' source of information in medicine and biology? *Physics in Medicine & Biology* **36**(1), 1.

Spence, G. R. (2011). *Directionally Sensitive Neutron Detector for Homeland Security Applications.*, Master's thesis, Texas A&M University, USA.

Stefanov, P. and Uhlmann, G. (2004). Stability estimates for the X-ray transform of tensor fields and boundary rigidity, *Duke Mathematical Journal* **123**(3), 445–467.

Stefanov, P. and Uhlmann, G. (2008). Integral geometry of tensor fields on a class of non-simple Riemannian manifolds, *American Journal of Mathematics* **130**(1), 239–268.

Stefanov, P., Uhlmann, G., and Vasy, A. (2018). Inverting the local geodesic X-ray transform on tensors, *Journal d'Analyse Mathematique* **136**(1), 151–208.

Sun, H. and Pistorius, S. (2014). Characterization the annihilation position distribution within a geometrical model associated with scattered coincidences in PET, in *2014 IEEE Nuclear Science Symposium and Medical Imaging Conference (NSS/MIC)*, 1–3.

Takeda, S., Harayama, A., Ichinohe, Y., Odaka, H., Watanabe, S., Takahashi, T., Tajima, H., Genba, K., Matsuura, D., Ikebuchi, H., Kuroda, Y., *et al.* (2015). A portable Si/CdTe Compton camera and its applications to the visualization of radioactive substances, *Nuclear Instruments and Methods in Physics Research Section A: Accelerators, Spectrometers, Detectors and Associated Equipment* **787**, 207–211.

Tan, H. and Gouia-Zarrad, R. (2017). New properties of the attenuated V-line transform for breast cancer detection with Compton cameras, *BIOMATH* **6**(2), 1711147.

Tarpau, C. and Nguyen, M. K. (2020). Compton scattering imaging system with two scanning configurations, *Journal of Electronic Imaging* **29**(1), 013005.

Tarpau, C., Cebeiro, J., Morvidone, M. A., and Nguyen, M. K. (2020a). A new concept of Compton scattering tomography and the development of the corresponding circular Radon transform, *IEEE Transactions on Radiation and Plasma Medical Sciences* **4**(4), 433–440.

Tarpau, C., Cebeiro, J., Nguyen, M. K., Rollet, G., and Dumas, L. (2020b). On the design of a CST system and its extension to a bi-imaging modality, *arXiv preprint arXiv:2007.02750* .

Tarpau, C., Cebeiro, J., Nguyen, M. K., Rollet, G., and Morvidone, M. A. (2020c). Analytic inversion of a Radon transform on double circular arcs with applications in Compton scattering tomography, *IEEE Transactions on Computational Imaging* **6**, 958–967.

Tashima, H., Yoshida, E., Wakizaka, H., Takahashi, M., Nagatsu, K., Tsuji, A. B., Kamada, K., Parodi, K., and Yamaya, T. (2020). 3D Compton image reconstruction method for whole gamma imaging, *Physics in Medicine & Biology* **65**(22), 225038.

Terzioglu, F. (2015). Some inversion formulas for the cone transform, *Inverse Problems* **31**(11), 115010.

Terzioglu, F. (2019). Some analytic properties of the cone transform, *Inverse Problems* **35**(3), 034002.

Terzioglu, F. (2020). Exact inversion of an integral transform arising in Compton camera imaging, *Journal of Medical Imaging* **7**, 3, p. 032504.

Terzioglu, F., Kuchment, P., and Kunyansky, L. (2018). Compton camera imaging and the cone transform: a brief overview, *Inverse Problems* **34**(5), 054002.

Todd, R. W., Nightingale, J. M., and Everett, D. B. (1974). A proposed γ camera, *Nature* **251**, 132–134, (Awarded the Ambrose Fleming Premium by the IEE in 1977).

Tomitani, T. and Hirasawa, M. (2002). Image reconstruction from limited angle Compton camera data, *Physics in Medicine & Biology* **47**(12), 2129–2145.

Towe, B. C. and Jacobs, A. M. (1981). X-ray backscatter imaging, *IEEE Transactions on Biomedical Engineering* **28**(9), 646–654.

Truong, T. (2013). Inversion of some spherical cap Radon transforms, *Eurasian Journal of Mathematical and Computer Applications* **1**(1), 78–102.

Truong, T. (2014). On geometric aspects of circular arcs Radon transforms for Compton scatter tomography, *Eurasian Journal of Mathematical and Computer Applications* **2**, 40–69.

Truong, T. and Nguyen, M. (2019). Compton scatter tomography in annular domains, *Inverse Problems* **35**(5), 054005.

Truong, T. T. and Nguyen, M. K. (2011a). On new V-line Radon transforms in $\mathbb{R}^2$ and their inversion, *Journal of Physics A: Mathematical and Theoretical* **44**(7), 075206.

Truong, T. T. and Nguyen, M. K. (2011b). Radon transforms on generalized Cormack's curves and a new Compton scatter tomography modality, *Inverse Problems* **27**(12), 125001.

Truong, T. and Nguyen, M. K. (2012). Recent developments on Compton scatter tomography: theory and numerical simulations, in M. Andriychuk (ed.), *Numerical Simulation-From Theory to Industry* (IntechOpen, Rijeka, Croatia), 101–128.

Truong, T. T. and Nguyen, M. K. (2015). New properties of the V-line Radon transform and their imaging applications, *Journal of Physics A: Mathematical and Theoretical* **48**(40), 405204.

Truong, T. T., Nguyen, M. K., and Zaidi, H. (2007). The mathematical foundations of 3D Compton scatter emission imaging, *International Journal of Biomedical Imaging* **2007**, p. 092780.

Tumer, T., Yin, S., and Kravis, S. (1997). A high sensitivity, electronically collimated gamma camera, *IEEE Transactions on Nuclear Science* **44**(3), 899–904.

Uche, C. Z., Round, W. H., and Cree, M. J. (2012). Evaluation of two Compton camera models for scintimammography, *Nuclear Instruments and Methods in Physics Research Section A: Accelerators, Spectrometers, Detectors and Associated Equipment* **662**(1), 55–60.

Valentine, J., Bonnerave, C., and Rohe, R. (1997). Energy-subtraction Compton scatter camera design considerations: a Monte Carlo study of timing and energy resolution effects, *IEEE Transactions on Nuclear Science* **44**(3), 1134–1139.

Van Rossum, M. C. W. and Nieuwenhuizen, T. M. (1999). Multiple scattering of classical waves: microscopy, mesoscopy, and diffusion, *Reviews of Modern Physics* **71**(1), 313–371.

Vertgeim, L. (2000). Integral geometry problems for symmetric tensor fields with incomplete data, *Journal of Inverse and Ill-posed Problems* **8**(3), 355–364.

Vinegoni, C., Pitsouli, C., Razansky, D., Perrimon, N., and Ntziachristos, V. (2008). In vivo imaging of Drosophila melanogaster pupae with mesoscopic fluorescence tomography, *Nature Methods* **9**, pp. 45–47.

Volchkov, V. V. (2003). *Integral Geometry and Convolution Equations* (Kluwer Academic Publishers, Dordrecht).

Wahl, C. G., Kaye, W. R., Wang, W., Zhang, F., Jaworski, J. M., King, A., Boucher, Y. A., and He, Z. (2015). The Polaris-H imaging spectrometer, *Nuclear Instruments and Methods in Physics Research Section A: Accelerators, Spectrometers, Detectors and Associated Equipment* **784**, 377–381.

Walker, M. R. and O'Sullivan, J. A. (2019). The broken ray transform: additional properties and new inversion formula, *Inverse Problems* **35**(11), 115003.

Walker, M. R. and O'Sullivan, J. A. (2021). Iterative algorithms for joint scatter and attenuation estimation from broken ray transform data, *IEEE Transactions on Computational Imaging* **7**, 361–374.

Wang, J., Chi, Z., and Wang, Y. (1999). Analytic reconstruction of Compton scattering tomography, *Journal of Applied Physics* **86**(3), 1693–1698.

Watanabe, S., Tanaka, T., Nakazawa, K., Mitani, T., Oonuki, K., Takahashi, T., Takashima, T., Tajima, H., Fukazawa, Y., Nomachi, M., *et al.* (2005). A Si/CdTe semiconductor Compton camera, *IEEE Transactions on Nuclear Science* **52**(5), 2045–2051.

Watson, C. C. (2000). New, faster, image-based scatter correction for 3D PET, *IEEE Transactions on Nuclear Science* **47**(4), 1587–1594.

Webber, C. and Kennett, T. (1976). Bone density measured by photon scattering. i. a system for clinical use, *Physics in Medicine & Biology* **21**(5), 760.

Webber, J. (2016). X-ray Compton scattering tomography, *Inverse Problems in Science and Engineering* **24**(8), 1323–1346.

Webber, J. and Miller, E. L. (2020). Compton scattering tomography in translational geometries, *Inverse Problems* **36**(2), 025007.

Webber, J. W. and Holman, S. (2019). Microlocal analysis of a spindle transform, *Inverse Problems & Imaging* **13**(2), 231–261.

Webber, J. W. and Quinto, E. T. (2020). Microlocal analysis of a Compton tomography problem, *SIAM Journal on Imaging Sciences* **13**(2), 746–774.

Webber, J. W. and Quinto, E. T. (2021). Microlocal analysis of generalized Radon transforms from scattering tomography, *SIAM Journal on Imaging Sciences* **14**(3), 976–1003.

Webber, J. W. and Lionheart, W. R. B. (2018). Three dimensional Compton scattering tomography, *Inverse Problems* **34**(8), 084001.

Webber, J. W., Quinto, E. T., and Miller, E. L. (2020). A joint reconstruction and lambda tomography regularization technique for energy-resolved X-ray imaging, *Inverse Problems* **36**(7), 074002.

Wernick, M. N. and Aarsvold, J. N. (2004). *Emission Tomography: The Fundamentals of PET and SPECT* (Elsevier Academic Press, Amsterdam).

Wightman, A. (1948). Note on polarization effects in Compton scattering, *Physical Review* **74**(12), 1813.

Wilderman, S., Rogers, W., Knoll, G., and Engdahl, J. (1997). Monte Carlo calculation of point-spread functions of Compton scatter cameras, *IEEE Transactions on Nuclear Science* **44**(2), 250–254.

Wilderman, S. J., Fessler, J. A., Clinthorne, N. H., LeBlanc, J., and Rogers, W. L. (2001). Improved modeling of system response in list mode EM reconstruction of Compton scatter camera images, *IEEE Transactions on Nuclear Science* **48**(1), 111–116.

Wilderman, S. J., Rogers, W., Knoll, G. F., and Engdahl, J. C. (1998). Fast algorithm for list mode back-projection of Compton scatter camera data, *IEEE Transactions on Nuclear Science* **45**(3), 957–962.

Wongsason, P. (2018). Vector field reconstruction via quaternionic setting, *Mathematical Methods in the Applied Sciences* **41**, 2, 684–696.

Wulf, E. A., Phlips, B. F., Johnson, W. N., Kurfess, J. D., and Novikova, E. I. (2004). Thick silicon strip detector Compton imager, *IEEE Transactions on Nuclear Science* **51**(5), 1997–2003.

Wulf, E. A., Phlips, B. F., Johnson, W. N., Kurfess, J. D., Novikova, E. I., O'Connor, P., and De Geronimo, G. (2007). Compton imager for detection of special nuclear material, *Nuclear Instruments and Methods in Physics Research Section A: Accelerators, Spectrometers, Detectors and Associated Equipment* **579**(1), 371–374.

Xu, D. and He, Z. (2006). Filtered back-projection in 4π Compton imaging with a single 3d position sensitive CdZnTe detector, *IEEE Transactions on Nuclear Science* **53**(5), 2787–2796.

Yagle, A. E. (1992). Inversion of spherical means using geometric inversion and Radon transforms, *Inverse Problems* **8**(6), 949–964.

Yang, Y., Gono, Y., Motomura, S., Enomoto, S., and Yano, Y. (2001). A Compton camera for multitracer imaging, *IEEE Transactions on Nuclear Science* **48**(3), 656–661.

Yuasa, T., Akiba, M., Takeda, T., Kazama, M., Hoshino, A., Watanabe, Y., Hyodo, K., Dilmanian, A., Akatsuka, T., and Itai, Y. (1997). Incoherent-scatter computed tomography with monochromatic synchrotron X-ray: feasibility of multi-CT imaging system for simultaneous measurement-of fluorescent and incoherent scatter X-rays, *IEEE Transactions on Nuclear Science* **44**(5), 1760–1769.

Zaidi, H. (1999). Relevance of accurate Monte Carlo modeling in nuclear medical imaging, *Medical Physics* **26**(4), 574–608.

Zaidi, H. and Koral, K. F. (2004). Scatter modelling and compensation in emission tomography, *European Journal of Nuclear Medicine and Molecular Imaging* **31**(5), 761–782.

Zhang, L., Rogers, W. L., and Clinthorne, N. H. (2004). Potential of a Compton camera for high performance scintimammography, *Physics in Medicine & Biology* **49**(4), 617.

Zhang, Y. (2020a). Artifacts in the inversion of the broken ray transform in the plane, *Inverse Problems & Imaging* **14**(1), 1.

Zhang, Y. (2020b). Recovery of singularities for the weighted cone transform appearing in Compton camera imaging, *Inverse Problems* **36**(2), 025014.

Zhao, F., Schotland, J. C., and Markel, V. A. (2014). Inversion of the star transform, *Inverse Problems* **30**(10), 105001.

Zoglauer, A. and Kanbach, G. (2003). Doppler broadening as a lower limit to the angular resolution of next-generation Compton telescopes, in J. E. Truemper and H. D. Tananbaum (eds.), *X-Ray and Gamma-Ray Telescopes and Instruments for Astronomy*, Vol. 4851, International Society for Optics and Photonics (SPIE), 1302–1309.

Index

www.ingramcontent.com/pod-product-compliance
Lightning Source LLC
Chambersburg PA
CBHW050555190326
41458CB00007B/2055